Glencoe McGraw-Hill

Math Triumphs

Book 2: Measurement, Geometry, and Algebra

Authors

Basich Whitney • Brown • Dawson • Gonsalves • Silbey • Vielhaber

 Glencoe

Photo Credits

All coins photographed by United States Mint.

All bills photographed by Michael Houghton/StudiOhio.

Cover Alamy; **iv** (1 7 8)File Photo, (2 3)The McGraw-Hill Companies, (4 5 6)Doug Martin; **vi** PunchStock; **vii** Robert Glusic/CORBIS; **viii** PunchStock; **140-141** (bkgd)Jim Lane/Alamy; **141** (inset)Ryan McVay/Getty Images; **142** Ken Cavanagh/The McGraw-Hill Companies; **143** Getty Images; **145** (bl)G.K. & Vikki Hart/Getty Images, (others)Alamy; **146** Jules Frazier/Getty Images; **147** Mark Ransom/RansomStudios; **148** (l)PunchStock, (r)CORBIS; **157, 159** PunchStock; **160** Getty Images; **161** PunchStock; **162, 169** Siede Preis/Getty Images; **172** (l)PunchStock, (r)Eric Misko/The McGraw-Hill Companies; **175** Jupiterimages; **178-179** Michael A. Keller/CORBIS; **184** (t)Getty Images, (b)Fotosearch; **191** (t)PunchStock, (b)Kevin Cavanagh/The McGraw-Hill Companies; **193** Getty Images; **200** Alamy; **204** Zigy Kaluzny/Getty Images; **205** iStockphoto; **206, 207** Sandra Ivany/Getty Images; **212** Getty Images; **214** PunchStock; **222** Helene Rogers/Alamy; **227** PunchStock; **229** CORBIS; **230** Getty Images; **240-241** Les Gibbon/Alamy; **245** PunchStock; **246** Ryan McVay/Getty Images; **252** Ken Cavanagh/The McGraw-Hill Companies; **253** Alamy; **255** Getty Images; **260** Judith Collins/Alamy; **262** Getty Images; **268** (t)Getty Images, (b)PunchStock; **269** GK & Vikki Hart/Getty Images; **275** Bob Pool/Getty Images.

The McGraw·Hill Companies

Macmillan/McGraw-Hill
Glencoe

Send all inquiries to:
Glencoe/McGraw-Hill
8787 Orion Place
Columbus, OH 43240-4027

ISBN: 978-0-07-888211-1
MHID: 0-07-888211-7

Math Triumphs
Grade 7, Book 2

Printed in the United States of America.

8 9 10 HSO 17 16 15 14 13 12

Math Triumphs

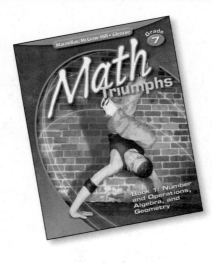

Book 1

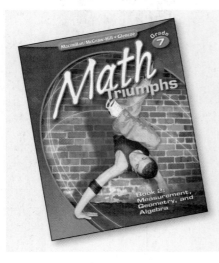

Book 2

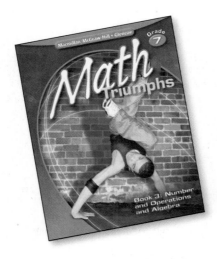

Book 3

Authors and Consultants

AUTHORS

Frances Basich Whitney
Project Director, Mathematics K–12
Santa Cruz County Office of Education
Capitola, California

Kathleen M. Brown
Math Curriculum Staff Developer
Washington Middle School
Long Beach, California

Dixie Dawson
Math Curriculum Leader
Long Beach Unified
Long Beach, California

Philip Gonsalves
Mathematics Coordinator
Alameda County Office of Education
Hayward, California

Robyn Silbey
Math Specialist
Montgomery County Public Schools
Gaithersburg, Maryland

Kathy Vielhaber
Mathematics Consultant
St. Louis, Missouri

CONTRIBUTING AUTHORS

Viken Hovsepian
Professor of Mathematics
Rio Hondo College
Whittier, California

FOLDABLES Study Organizer **Dinah Zike**
Educational Consultant,
Dinah-Might Activities, Inc.
San Antonio, Texas

CONSULTANTS

Assessment

Donna M. Kopenski, Ed.D.
Math Coordinator K–5
City Heights Educational Collaborative
San Diego, California

Instructional Planning and Support

Beatrice Luchin
Mathematics Consultant
League City, Texas

ELL Support and Vocabulary

ReLeah Cossett Lent
Author/Educational Consultant
Alford, Florida

Reviewers

Each person below reviewed at least two chapters of the Student Edition, providing feedback and suggestions for improving the effectiveness of the mathematics instruction.

Patricia Allanson
Mathematics Teacher
Deltona Middle School
Deltona, Florida

Amy L. Chazarreta
Math Teacher
Wayside Middle School
Fort Worth, Texas

David E. Ewing
Teacher
Bellview Middle School
Pensacola, Florida

Russ Lush
Sixth Grade Math Teacher/Math Dept. Chair
New Augusta - North
Indianapolis, Indiana

Karen L. Reed
Sixth Pre-AP Math
Keller ISD
Keller, Texas

Debra Allred
Sixth Grade Math Teacher
Wiley Middle School
Leander, Texas

Jeff Denney
Seventh Grade Math Teacher, Mathematics
   Department Chair
Oak Mountain Middle School
Birmingham, Alabama

Mark J. Forzley
Eighth Grade Math Teacher
Westmont Junior High School
Westmont, Illinois

Joyce B. McClain
Middle School Math Consultant
Hillsborough County Schools
Tampa, Florida

Deborah Todd
Sixth Grade Math Teacher
Francis Bradley Middle School
Huntersville, North Carolina

April Chauvette
Secondary Mathematics Facilitator
Leander Independent School District
Leander, Texas

Franco A. DiPasqua
Director of K-12 Mathematics
West Seneca Central
West Seneca, New York

Virginia Granstrand Harrell
Education Consultant
Tampa, Florida

Suzanne D. Obuchowski
Math Teacher
Proctor School
Topsfield, Massachusetts

Susan S. Wesson
Teacher (retired)
Pilot Butte Middle School
Bend, Oregon

Contents

Chapter 4 — Measurement

Waipio Valley, Hawaii

Contents

Chapter 5

Two-Dimensional Figures

Chicago, Illinois

Contents

Chapter 6 — Three-Dimensional Figures

Bucks County, Pennsylvania

SCAVENGER HUNT

BOOK 2

Let's Get Started

Use the Scavenger Hunt below to learn where things are located in each chapter.

1. What is the title of Lesson 5-2?

2. What is the Key Concept of Lesson 4-4?

3. On what page can you find the vocabulary term *customary system* in Lesson 4-3?

4. What are the vocabulary words for Lesson 6-2?

5. How many Examples are presented in the Chapter 4 Study Guide?

6. What strategy is used in the Step-by-Step Problem-Solving Practice box on page 245?

7. List the measurements that are mentioned in exercise #13 on page 154.

8. Describe the art that accompanies exercise #6 on page 145?

9. On what pages will you find the Test Practice for Chapter 6?

10. In Chapter 5, find the logo and Internet address that tells you where you can take the Online Readiness Quiz.

Chapter 4

Measurement

Why do we use measurement?

We use measurements every day. A doctor measures how tall a patient is, a carpenter measures lengths of wood to build a house, and the distance in a race can be measured in meters. We measure objects using metric and customary units of length.

Step 1 Quiz

Math Online ⟩ **Are you ready for Chapter 4? Take the Online Readiness Quiz at** *glencoe.com* **to find out.**

Step 2 Preview

Get ready for Chapter 4. Review these skills and compare them with what you will learn in this chapter.

What You Know	What You Will Learn
You know how to multiply and divide by powers of ten.	*Lessons 4-1 and 4-2*

What You Know

You know how to multiply and divide by powers of ten.

Examples: $4 \times 1{,}000 = 4{,}000$
$300 \div 100 = 3$

TRY IT!

1 $5 \times 10 =$ _____

2 $7 \times 100 =$ _____

3 $12 \times 1{,}000 =$ _____

4 $90 \div 10 =$ _____

5 $24 \div 100 =$ _____

6 $15 \div 1{,}000 =$ _____

What You Will Learn

Lessons 4-1 and 4-2

The **metric system** is a measurement system in which units differ from the base unit by a power of ten.

1 kilometer = 1,000 meters

So, 4 kilometers = $4 \times 1{,}000$, or 4,000 meters.

1 meter = 0.001 kilometers

So, 4 meters = $4 \div 1{,}000$ or 0.004 kilometers.

You know how to multiply and divide.

Examples: $3 \times 12 = 36$
$15 \div 3 = 5$

TRY IT!

7 $12 \times 4 =$ _____

8 $36 \times 3 =$ _____

9 $1{,}760 \times 5 =$ _____

10 $60 \div 12 =$ _____

11 $93 \div 3 =$ _____

12 $7{,}040 \div 1{,}760 =$ _____

Lessons 4-3 and 4-4

The **customary system** of measurement uses units such as foot and yard. You multiply or divide to change units.

1 foot = 12 inches
So, 3 feet = 3×12, or 36 inches.

3 feet = 1 yard
So, 30 feet = $30 \div 3$, or 10 yards.

Metric Length

KEY Concept

The metric system is a decimal system of weights and measures. A **meter** is the base unit of length in the **metric system**.

The most commonly used metric units of **length** are shown below.

Metric Units of Length			
Metric Unit	Symbol	Real-World Benchmark	Meaning
millimeter	mm	thickness of a dime	one-thousandth
centimeter	cm	half the width of a penny	one-hundredth
meter	m	height of a doorknob	one
kilometer	km	six city blocks	one-thousand

Units on a centimeter ruler are divided into ten parts. Each part is a millimeter.

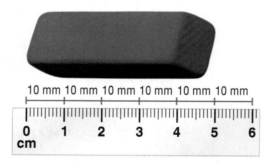

To read millimeters, count each individual mark on the centimeter ruler. There are ten millimeter marks for each centimeter mark.

The eraser is about 5 centimeters or 50 millimeters long.

VOCABULARY

centimeter
a metric unit of length; one centimeter equals one-hundredth of a meter

kilometer
a metric unit of length; one kilometer equals one thousand meters

length
a measurement of the distance between two points

meter
the base unit of length in the metric system; one meter equals one-thousandth of a kilometer

metric system
a decimal system of weights and measures

millimeter
a metric unit of length; one millimeter equals one-thousandth of a meter

Example 1

Find the length of the golf tee to the nearest centimeter.

1. Line up the "zero mark" of a ruler with the left end of the golf tee.

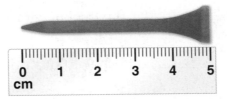

2. Read the number on the ruler that lines up with the right end of the golf tee.

The golf tee is 5 centimeters long.

YOUR TURN!

Find the length of the paper clip to the nearest centimeter.

1. Line up the "zero mark" of a ruler with the left end of the paper clip.

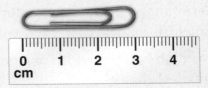

2. Read the number on the ruler that lines up with the right end of the paper clip.

The paper clip is about _____ centimeters long.

Example 2

Find the length of the ticket to the nearest millimeter.

1. Line up the "zero mark" of the ruler with the left end of the ticket.

2. Read the mark on the ruler that lines up with the right end of the ticket by counting each individual mark.

The ticket is about 59 millimeters long.

YOUR TURN!

Find the length of the quarter to the nearest millimeter.

1. Line up the "zero mark" of a ruler with the left end of the quarter.

2. Read the mark on the ruler that lines up with the right end of the quarter by counting each individual mark.

The quarter is about _____ millimeters long.

GO ON

Example 3

Which unit would you use to measure the height of your school: *millimeter*, *centimeter*, *meter*, or *kilometer*?

The height of your school is . . .

1. **greater than** the thickness of a dime. (millimeter)

2. **greater than** half the width of a penny. (centimeter)

3. **less than** six city blocks. (kilometer)

The **meter** is an appropriate unit of measure.

YOUR TURN!

Which unit would you use to measure the width of a skateboard: *millimeter*, *centimeter*, *meter*, or *kilometer*?

The width of a skateboard is . . .

1. _____ the thickness of a dime.

2. _____ the height of a doorknob.

3. _____ six city blocks.

The _____ is an appropriate unit of measure.

Who is Correct?

Find the length of the yarn to the nearest millimeter.

Lucy	Brian	Adina
60 millimeters	**57 millimeters**	**50 millimeters**

Circle correct answer(s). Cross out incorrect answer(s).

 Guided Practice

Draw a line segment of each length.

1 10 centimeters

2 48 millimeters

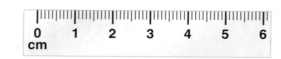

3 Find the width of the golf ball to the nearest centimeter.

Step 1 Line up the "zero mark" of the ruler with the _____ of the golf ball.

Step 2 Read the number on the ruler that lines up with the _____ of the golf ball.

The golf ball is about _____ centimeters wide.

Find the length of each line segment.

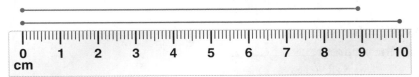

4 Red line segment to the nearest millimeter: _____

5 Blue line segment to the nearest centimeter: _____

Find the length of each object to the nearest centimeter or millimeter.

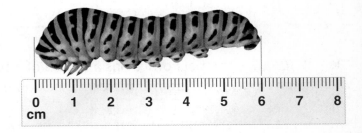

6 _____

7 _____

GO ON

Select the proper unit to measure the length of each object. Write *millimeter, centimeter, meter,* **or** *kilometer.*

8 thickness of a calculator _____

9 width of a notebook _____

10 height of a flag pole _____

11 distance from home to school _____

Step by Step **Problem-Solving Practice**

Solve.

12 BASEBALL Gloria has a model of a baseball bat. What is the length of the model baseball bat to the nearest centimeter?

Problem-Solving Strategies
- ☐ Look for a pattern.
- ☐ Guess and check.
- ☑ Act it out.
- ☐ Solve a simpler problem.
- ☐ Work backward.

Understand	Read the problem. Write what you know. Measure the length to the nearest _____.
Plan	Pick a strategy. One strategy is to act it out. Line up the 0 on a centimeter ruler with the bat.
Solve	Read the closest number on the ruler that lines up with the right end of the baseball bat. The baseball bat is about _____ centimeters.
Check	The baseball bat is greater than 11 centimeters and less than 12 centimeters long. The answer makes sense.

13 MODELS The post office has a model of a flagpole. What is the height to the nearest millimeter? Check off each step.

_____ **Understand: I underlined the key words.**

_____ **Plan: To solve the problem, I will** _____.

_____ **Solve: The answer is** _____.

_____ **Check: I checked my answer by** _____.

14 SHOPPING   Darcy bought a ribbon for a dress. What is the length of the ribbon to the nearest centimeter? _____

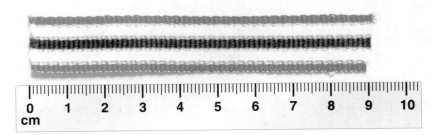

15 Reflect How would you explain to someone how to measure the length of a cell phone?

_____

_____

_____

▶ Skills, Concepts, and Problem Solving

Draw a line segment of each length.

16 4 centimeters

17 95 millimeters

Find the length of each line segment.

18 To the nearest millimeter:

_____

19 To the nearest centimeter:

_____

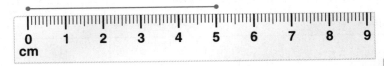

GO ON

Find the length of each object to the nearest centimeter or millimeter.

20 _____

21 _____

Write the metric unit of length that you would use to measure each of the following.

22 distance between two cities _____

23 length of a ladybug _____

24 width of a baseball card _____

25 height of a house _____

Solve.

26 SCHOOL After Seth sharpened his pencil, it was 101 millimeters long. Draw a line segment to show the length of Seth's pencil.

27 SCIENCE A snail traveled 5 centimeters in an hour. Draw a line segment to show this distance.

Vocabulary Check **Write the vocabulary word that completes each sentence.**

28 One _____ equals one thousand meters.

29 The _____ is the base unit of length in the metric system.

30 Writing in Math Carlie said she was going to measure the length of her bedroom in millimeters. Is this a good choice? Explain.

_____

_____

STOP

Unit Conversions: Metric Length

KEY Concept

Prefixes used for units of metric measurement always have the same meaning. The **meter** is the base unit of length in the **metric system**. Each prefix shows the size of a unit compared to a meter.

Prefix	Meaning	Metric Unit	Symbol	Real-World Benchmark
milli	one-thousandth	millimeter	mm	thickness of a dime
centi	one-hundredth	centimeter	cm	half the width of a penny
deci	one-tenth	decimeter	dm	length of a crayon
	one	meter	m	height of a doorknob
kilo	one thousand	kilometer	km	six city blocks

Sometimes it is necessary to **convert** from one unit of measurement to another. A metric place-value chart can be useful.

1000	100	10	1	0.1	0.01	0.001
thousands	hundreds	tens	ones	tenths	hundredths	thousandths
kilo (km)			meters (m)	deci (dm)	centi (cm)	milli (mm)

The following metric conversion diagram can also be used to help convert metric units of measure. To convert a larger unit to a smaller unit, you should multiply. To convert a smaller unit to a larger unit, you should divide.

$\times 1000 \quad \times 100 \quad \times 10$

larger units → km ⤻ m ⤻ cm ⤻ mm ← smaller units

$\div 1000 \quad \div 100 \quad \div 10$

VOCABULARY

benchmark
an object or number used as a guide to estimate or reference

centimeter
a metric unit of length; one centimeter equals one-hundredth of a meter

convert
to find an equivalent measure

kilometer
a metric unit of length; one kilometer equals one thousand meters

meter
the base unit of length in the metric system; one meter equals one-thousandth of a kilometer

metric system
a decimal system of weights and measures

millimeter
a metric unit of length; one millimeter equals one-thousandth of a meter

GO ON

Example 1

Convert 8 centimeters to meters.

1. Use a chart. Place 8 in the cm column.

1000				1	0.1	0.01	0.001
thousands				ones	tenths	hundredths	thousandths
				O	O	8	
kilo (km)				meters (m)	deci (dm)	centi (cm)	milli (mm)

> The chart is set up this way because a centimeter is $\frac{1}{100}$ of a meter. A decimeter is $\frac{1}{10}$ of a meter.

2. Place zeros in the m and dm columns.

3. Read the number from the chart for the conversion. $8 \text{ cm} = 0.08 \text{ m}$

YOUR TURN!

Convert 9.7 decimeters to meters.

1. Use a chart. Place _____ in the dm column and _____ in the cm column.

1000				1	0.1	0.01	0.001
thousands				ones	tenths	hundredths	thousandths
kilo (km)				meters (m)	deci (dm)	centi (cm)	milli (mm)

2. Place a zero in the _____ column.

3. Read the number from the chart for the conversion.

$9.7 \text{ dm} = $ _____ m

Example 2

Convert 6.4 kilometers to meters.

1. Use a chart. Place 6 in the km column and 4 in the next column to the right.

1000				1	0.1	0.01	0.001
thousands				ones	tenths	hundredths	thousandths
6	4	O	O				
kilo (km)				meters (m)	deci (dm)	centi (cm)	milli (mm)

2. Place zeros in the columns between 4 and the decimal point.

3. Read the number from the chart for the conversion. $6.4 \text{ km} = 6{,}400 \text{ m}$

YOUR TURN!

Convert 3 meters to millimeters.

1. Use a chart. Place _____ in the m column.

1000				1	0.1	0.01	0.001
thousands				ones	tenths	hundredths	thousandths
kilo (km)				meters (m)	deci (dm)	centi (cm)	milli (mm)

2. Place zeros in the dm, cm, and mm columns.

3. Read the number from the chart for the conversion.

$3 \text{ m} = $ _____ mm

Example 3

Convert 3 meters to centimeters.

1. **1** meter is equal to 100 centimeters.

2. You are converting from a **larger** unit to a **smaller** unit, so **multiply**.

3. Convert.

$3 \times 100 = 300$

$3 \text{ m} = 300 \text{ cm}$

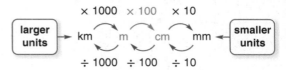

YOUR TURN!

Convert 7 meters to kilometers.

1. _____ m is equal to 1 kilometer.

2. You are converting from a _____ unit to a _____ unit, so _____.

3. Convert.

7 _____ = _____

$7 \text{ m} =$ _____ km

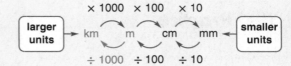

Who is Correct?

Convert 8.4 meters to millimeters.

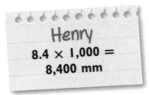

Clara
$8.4 \times 10{,}000 =$
84,000 mm

Andre
$8.4 \div 1{,}000 =$
0.0084 mm

Henry
$8.4 \times 1{,}000 =$
8,400 mm

Circle correct answer(s). Cross out incorrect answer(s).

GO ON

▶ Guided Practice

Convert using a place-value chart.

1   7 km = _____ m

1000			1	0.1	0.01	0.001
thousands			ones	tenths	hundredths	thousandths
kilo (km)			meters (m)	deci (dm)	centi (cm)	milli (mm)

2   3 dm = _____ m

1000			1	0.1	0.01	0.001
thousands			ones	tenths	hundredths	thousandths
kilo (km)			meters (m)	deci (dm)	centi (cm)	milli (mm)

Step by Step Practice

Convert.

3   8 m = _____ mm

     Step 1   _____ millimeters is equal to 1 meter.

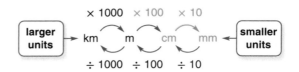

     Step 2   You are converting from a _____ unit to a

                 _____ unit, so you _____.

     Step 3   Convert.   8 _____ = _____

                 8 m = _____ mm

Convert.

4   5 m = _____ cm

     1 m = _____ cm

     Multiply or divide? _____

     5 _____ 100 = _____

     5 m = _____ cm

5   8 m = _____ km

     1 km = _____ m

     Multiply or divide? _____

     8 _____ 1,000 = _____

     8 m = _____ km

6   8.5 cm = _____ m

7   0.5 m = _____ dm

8   93 dm = _____ m

9   2 km = _____ m

Step by Step Problem-Solving Practice

Solve.

Problem-Solving Strategies
☐ Draw a diagram.
☑ Look for a pattern.
☐ Act it out.
☐ Solve a simpler problem.
☐ Work backward.

10 SPORTS   A soccer field is 120 meters long. How many decimeters long is a soccer field?

Understand   Read the question. Write what you know.

A soccer field is _____ meters long.

Plan   Pick a strategy. One strategy is to look for a pattern.

_____ decimeters is equal to 1 meter. Find a

rule. One rule is to add _____ for each meter.

Solve   The pattern is to add 10, 120 times. This is repeated

addition, which is the same as _____ × _____.

The soccer field is _____ decimeters long.

Check   Think: Decimeters are a smaller unit of measure than meters, so the number of decimeters of a soccer field is greater than the number of meters.

---

11 SEWING   Booker bought 1,850 millimeters of ribbon to make a pillow. The pillow required 170 centimeters of ribbon. In centimeters, how much extra ribbon is left?

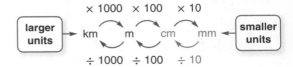

Check off each step.

_____ **Understand: I underlined key words.**

_____ **Plan: To solve the problem, I will** _____.

_____ **Solve: The answer is** _____.

_____ **Check: I checked my answer by** _____.

12 SHOES The sales clerk measured Robert's foot to be 3.2 decimeters long. How many millimeters long is Robert's foot?

_____

13 Reflect Is 700 millimeters equal to 7 meters? Use patterns to explain.

_____

▶ Skills, Concepts, and Problem Solving

Convert using a place-value chart.

14 7 cm = _____ m

1000			1	0.1	0.01	0.001
thousands			ones	tenths	hundredths	thousandths
kilo (km)			meters (m)	deci (dm)	centi (cm)	milli (mm)

15 6 km = _____ m

1000			1	0.1	0.01	0.001
thousands			ones	tenths	hundredths	thousandths
kilo (km)			meters (m)	deci (dm)	centi (cm)	milli (mm)

16 73 dm = _____ cm

1 dm = _____ cm

Multiply or divide? _____

73 _____ 10 = _____

73 dm = _____ cm

17 6 m = _____ mm

1 m = _____ mm

Multiply or divide? _____

_____ × 1,000 = _____

6 m = _____ mm

Convert.

18 70 m = _____ dm

19 58.6 cm = _____ m

20 92.7 mm = _____ cm

21 360 m = _____ km

22 4.3 m = _____ mm

23 0.021 km = _____ m

24 4.9 dm = _____ cm

25 6.4 cm = _____ m

Solve.

26 **TRAVEL**   It is 63 kilometers from Grady's house to his cousin's house. How many meters is it to Grady's cousin's house?

_____

27 **PETS**   Marni's cat was found wandering around the high school which is 4,700 meters from her home. How many kilometers away was Marni's cat? _____

Vocabulary Check   **Write the vocabulary word that completes each sentence.**

28 The _____ system is a decimal system of weights and measures.

29 A _____ is the base unit of length in the metric system.

30 **Writing in Math**   Explain how to convert 5.2 meters to centimeters.

_____

_____

▶ **Spiral Review** (Lesson 4-1 p. 142)

Find the length of the line segment to the nearest millimeter or centimeter.

31 _____ cm

32 _____ mm

Select the proper unit to measure the length of each object. Write millimeter, centimeter, meter, or kilometer.

33 length of an ant _____

34 height of a giraffe _____

35 **SWIMMING**   Which is the most appropriate estimate for the depth of a swimming pool: 3 millimeters, 3 meters, or 3 kilometers? Explain.

_____

_____

STOP

Progress Check 1 (Lessons 4-1 and 4-2)

Draw a line segment of each length.

1 52 millimeters

0	1	2	3	4	5	6
cm						

2 7 centimeters

0	1	2	3	4	5	6	7	8
cm								

Find the length of each line segment to the nearest millimeter or centimeter.

3 The line segment is about _____ millimeters long.

0	1	2	3	4	5	6	7	8	9
cm									

4 The line segment is about _____ centimeters long.

0	1	2	3	4	5	6
cm						

Convert using a place-value chart.

5 73 km = _____ m

10,000	1000			1	0.1	0.01	0.001
ten thousands	thousands			ones	tenths	hundredths	thousandths
	kilo (km)			meters (m)	deci (dm)	centi (cm)	milli (mm)

6 45 mm = _____ m

1000			1	0.1	0.01	0.001
thousands			ones	tenths	hundredths	thousandths
kilo (km)			meters (m)	deci (dm)	centi (cm)	milli (mm)

Convert.

7 0.68 cm = _____ dm

8 103 mm = _____ m

9 439 dm = _____ mm

10 1,800 m = _____ km

Solve.

11 **LAND** The road on Rachel's farm is 2,475 meters long. How many kilometers long is the road? _____

Customary Length

KEY Concept

The units of **length** most often used in the United States are the inch, foot, yard, and mile. These units are part of the customary system.

Units of Length

Customary Unit	Symbol	Real-World Benchmark
1 inch	in.	width of a quarter
1 foot	ft	length of a large adult foot
1 yard	yd	length from nose to fingertip
1 mile	mi	four laps around a running track

You can use a ruler to measure objects to the nearest half inch or quarter inch.

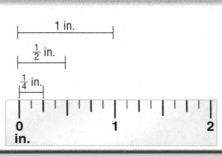

1 in.

$\frac{1}{2}$ in.

$\frac{1}{4}$ in.

VOCABULARY

customary system
a measurement system that includes units such as foot, pound, and quart

foot
a customary unit of length equal to 12 inches

inch
a customary unit of length; 12 inches equal 1 foot

length
a measurement of the distance between two points

mile
a customary unit of length equal to 5,280 feet or 1,760 yards

yard
a customary unit of length equal to 3 feet, or 36 inches

Example 1

Find the length of the nail to the nearest $\frac{1}{2}$ inch.

1. Line up the "zero mark" of the ruler with the left end of the nail.

2. Find the $\frac{1}{2}$ inch mark that is closest to the right end.

The nail is $2\frac{1}{2}$ inches long.

YOUR TURN!

Find the length of the grasshopper to the nearest $\frac{1}{2}$ inch.

1. Line up the "zero mark" of a ruler with the left end of the grasshopper.

2. Find the $\frac{1}{2}$ inch mark that is closest to the right end.

The grasshopper is _____ inches long.

GO ON

Example 2

Find the length of the pencil to the nearest $\frac{1}{4}$ inch.

1. Line up the "zero mark" of the ruler with the left end of the pencil.

2. Find the $\frac{1}{4}$ inch mark that is closest to the right end.

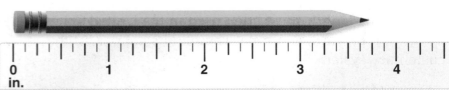

The pencil is about $3\frac{3}{4}$ inches long.

YOUR TURN!

Find the length of the pen to the nearest $\frac{1}{4}$ inch.

1. Line up the "zero mark" of the ruler with the left end of the pen.

2. Find the $\frac{1}{4}$ inch mark that is closest to the right end.

The pen is about _____ inches long.

Example 3

Which unit would you use to measure the length of a bicycle: *inch*, *foot*, *yard*, or *mile*?

The length of a bicycle is . . .

1. **greater than** the width of a quarter.

2. **less than** the length from nose to fingertip.

3. **less than** four laps around a running track.

The **foot** is an appropriate unit of measure.

YOUR TURN!

Which unit would you use to measure the length of a football field: *inch*, *foot*, *yard*, or *mile*?

The length of a football field is . . .

1. _____ the width of a quarter.

2. _____ the length of a large adult foot.

3. _____ four laps around a running track.

The _____ is an appropriate unit of measure.

Who is Correct?

Find the length of the chalk to the nearest $\frac{1}{4}$ inch.

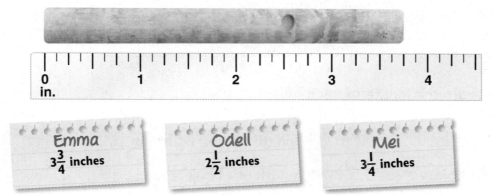

Emma
$3\frac{3}{4}$ inches

Odell
$2\frac{1}{2}$ inches

Mei
$3\frac{1}{4}$ inches

Circle correct answer(s). Cross out incorrect answer(s).

▶ Guided Practice

Draw a line segment of each length.

1 $5\frac{1}{4}$ inches

2 $3\frac{3}{4}$ inches

Step (by) Step Practice

3 Find the length of the eraser to the nearest $\frac{1}{4}$ inch.

Step 1 Line up the "zero mark" of the ruler with the
_____ of the eraser.

Step 2 Find the $\frac{1}{4}$ inch mark that is closest to the
_____ of the eraser.

The eraser is about _____ inches long.

GO ON

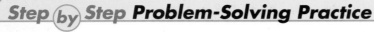

4 Measure the length of the red line segment to the nearest $\frac{1}{4}$ inch. _____

5 Measure the length of the blue line segment to the nearest $\frac{1}{2}$ inch. _____

Select the proper unit to measure the length of each object. Write inch, foot, yard, or mile.

6 height of your teacher _____

7 length of your bedroom _____

8 distance from school to the library _____

9 width of a book _____

Step (by) Step *Problem-Solving Practice*

Solve.

10 **ART**   Carlos has a pea pod. What is the length of the pea pod to the nearest $\frac{1}{2}$ inch?

<table>
<tr><td>Problem-Solving Strategies</td></tr>
<tr><td>☐ Look for a pattern.</td></tr>
<tr><td>☐ Guess and check.</td></tr>
<tr><td>☑ Act it out.</td></tr>
<tr><td>☐ Solve a simpler problem.</td></tr>
<tr><td>☐ Work backward.</td></tr>
</table>

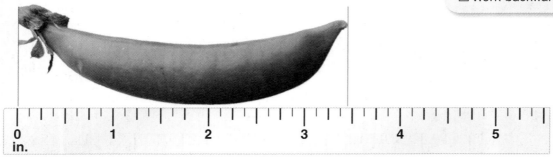

Understand   Read the problem. Write what you know.

Measure the length to the nearest _____.

Plan   Pick a strategy. One strategy is to act it out. Line up the zero mark on an inch ruler with the end of the pea pod.

Solve   Find the half-inch mark that is closest to the right end of the pea pod.

The pea pod is about _____ inches long.

Check   The pea pod is greater than 3 and less than 4 inches long. The answer makes sense.

11 **SCIENCE** Ryan is collecting shells to make a craft. What is the length of the shell shown to the nearest $\frac{1}{4}$ inch? Check off each step.

_____ Understand: I underlined the key words.

_____ Plan: To solve the problem, I will _____.

_____ Solve: The answer is _____.

_____ Check: I checked my answer by _____.

12 **WOODWORKING** Maria is using screws to build a doghouse. What is the length of the screws to the nearest $\frac{1}{4}$ inch?

_____

13 **Reflect** How would you explain to someone what customary system unit to use when measuring the distance from Tampa, Florida, to Miami, Florida?

_____

_____

▶ Skills, Concepts, and Problem Solving

Draw a line segment of each length.

14 $5\frac{1}{2}$ inches

15 $4\frac{1}{4}$ inches

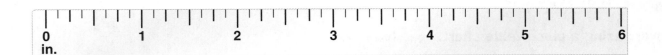

Measure the length of each line segment to the nearest $\frac{1}{4}$ or $\frac{1}{2}$ inch.

16 Red line segment to the nearest $\frac{1}{2}$ inch:

17 Blue line segment to the nearest $\frac{1}{4}$ inch:

_____   _____

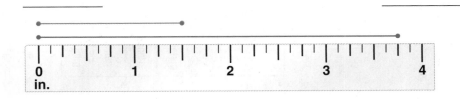

GO ON

Find the length of each object.

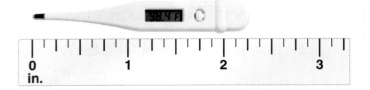

18 _____ in.

19 _____ in.

Solve.

20 **TECHNOLOGY** The screen on Tamika's new video player measured $3\frac{1}{2}$ inches long. Draw a line segment to show its width.

Vocabulary Check **Write the vocabulary word that completes each sentence.**

21 A(n) _____ is a unit of length equal to 5,280 feet or 1,760 yards.

22 A(n) _____ is a unit of length equal to 12 inches.

23 **Writing in Math** Explain how to measure a line segment to the nearest $\frac{1}{2}$ inch.

_____

_____

▶ **Spiral Review**

Convert using a place-value chart. (Lesson 4-2, p. 149)

24 3 m = _____ cm

1000				1	0.1	0.01	0.001
thousands				ones	tenths	hundredths	thousandths
				•			
kilo (km)				meters (m)	deci (dm)	centi (cm)	milli (mm)

25 9 m = _____ km

1000				1	0.1	0.01	0.001
thousands				ones	tenths	hundredths	thousandths
				•			
kilo (km)				meters (m)	deci (dm)	centi (cm)	milli (mm)

STOP

Unit Conversions: Customary Length

KEY Concept

Unit for Length	Abbreviation	Equivalents	Real-World Benchmark
inch	in.		small paper clip
foot	ft	1 ft = 12 in.	standard ruler
yard	yd	1 yd = 3 ft 1 yd = 36 in.	baseball bat
mile	mi	1 mi = 1,760 yd 1 mi = 5,280 ft	about eight city blocks

Use the last column of the table to help you understand the relative size of a unit by comparing it to everyday objects.

Use a ruler to see how the units of length compare.

Sometimes it is necessary to **convert** from one unit of measure to another. Knowing customary conversions can help you understand the relationship between two units.

Copyright © Glencoe/McGraw-Hill, a division of The McGraw-Hill Companies, Inc.

VOCABULARY

benchmark
an object or number used as a guide to estimate or reference

convert
to find an equivalent measure

customary system
a measurement system that includes units such as foot, pound, and quart

foot
a customary unit of length equal to 12 inches

inch
a customary unit of length; 12 inches equal 1 foot

mile
a customary unit of length equal to 5,280 feet or 1,760 yards

yard
a customary unit of length equal to 3 feet or 36 inches

Example 1

Convert 60 inches to feet using a table.

feet	1	2	3	4	5
inches	12	24	36	48	60

1. There are 12 inches in 1 foot.
2. Fill in the table.

   2 feet = 2 × 12 inches
   3 feet = 3 × 12 inches
   4 feet = 4 × 12 inches
   5 feet = 5 × 12 inches

60 inches are equal to 5 feet.

YOUR TURN!

Convert 15 feet to yards using a table.

yards					
feet	3	6	9	12	15

1. There are 3 feet in 1 yard. Complete the chart by using multiples of three.
2. Fill in the table.

   _____ feet are equal to 5 yards.

To convert a larger unit to a smaller unit, multiply.
To convert a smaller to a larger unit, divide.

Example 2

Convert 2.5 feet to inches.

1. You are converting from feet to inches, which is a larger unit to a smaller unit. You should multiply.

2. 1 foot is equal to 12 inches.

So, 2.5 feet are equal to 2.5 × 12, or 30 inches.

YOUR TURN!

Convert 156 feet to yards.

1. You are converting from feet to yards, which is a smaller unit to a larger unit. You should _____.

2. _____ feet are equal to _____ yard.

So, 156 feet are equal to

_____ ÷ _____ ,

or _____ yards.

Who is Correct?

Convert 48 inches to feet.

Che
12 inches are equal
to 1 foot.
48 × 12 = 576 feet

Lucita
12 inches are equal
to 1 foot.
48 ÷ 12 = 4 feet

Graham
3 feet are equal to
1 yard and 12 inches
are equal to 1 foot.
48 ÷ 3 = 16 feet

Circle correct answer(s). Cross out incorrect answer(s).

▶ Guided Practice

Convert using a table.

1 6 yd = _____ in.

yards	1	2	3	4	5	6
inches						

2 4 mi = _____ yd

miles	1	2	3	4
yards				

Step by Step Practice

Convert.

3 9 yd = _____ ft

    Step 1 You are converting from a _____ unit to a
                   _____ unit, so you should _____.

    Step 2 1 yard is equal to _____ feet.

    Step 3 So, 9 yards are 9 _____ 3, or _____ feet.

Convert.

4 2 mi = _____ ft

    1 mi = _____ ft

    Multiply or divide? _____

    2 _____ 5,280 = _____

    2 mi = _____ ft

5 72 in. = _____ ft

    1 ft = _____ in.

    Multiply or divide? _____

    72 _____ 12 = _____

    72 in. = _____ ft

6 8,800 yd = _____ mi

7 3 mi = _____ yd

8 5 yd = _____ ft

9 5 mi = _____ ft

10 9 yd = _____ ft

11 1 mi = _____ yd

12 60 in. = _____ ft

13 15 ft = _____ in.

14 9 ft = _____ in.

15 12 ft = _____ in.

16 108 in. = _____ yd

17 93 ft = _____ yd

GO ON

Step by Step Problem-Solving Practice

Solve.

18 **HOMES**   The bedroom in Tenisha's apartment is 144 inches long. How many yards long is the room?

Understand   Read the question. Write what you know.
The bedroom is _____ inches long.

Plan   Pick a strategy. One strategy is to work backward.

You know the total number of inches.
Subtract repeatedly until the answer is 0.
Count the number of times you subtracted 36.

Solve   144   − 36 = _____   _____ yard

_____ − 36 = _____   _____ yards

_____ − 36 = _____   _____ yards

_____ − 36 = _____   _____ yards

The room is _____ yards long.

Check   Think: An inch is a smaller unit of measure than a yard. So the number of inches should be greater than the number of yards. The answer makes sense.

19 **SCHOOL**   Justina's desk is 42 inches wide. How many feet wide is her desk? Check off each step.

_____ Understand: I underlined the words.

_____ Plan: To solve the problem, I will _____.

_____ Solve: The answer is _____.

_____ Check: I checked my answer by _____.

20 **SPORTS**   During Saturday's football game, James set the school record by running 96 yards to score a touchdown. How many feet did James run for the touchdown?

_____

21 **Reflect** Is 108 inches equal to 9 feet? Explain.

_____

_____

▶ Skills, Concepts, and Problem Solving

Convert using a table.

22 8 ft = _____ in.

feet	1	2	3	4	5	6	7	8
inches								

23 4 yd = _____ in.

yards	1	2	3	4
inches				

Convert.

24 2 mi = _____ ft

25 39 ft = _____ yd

26 26,400 ft = _____ mi

27 10 mi = _____ ft

28 360 in. = _____ yd

29 17,600 yd = _____ mi

30 1,821 ft = _____ yd

31 17 yd = _____ in.

32 45 ft = _____ yd

33 2.5 yd = _____ in.

34 Ruben made a toy chest for his little sister. What are the dimensions in inches?

1 ft = _____ in.

1.5 ft = _____ in.

2 ft = _____ in.

GO ON

Solve.

35 **HISTORY** One of the largest balls of string is in Branson, Missouri. How many inches is the circumference of the ball of string?

_____

41.5 ft

36 **DECORATING** Olivia is redecorating her bedroom. She measured the length as 138 inches. She measured the width as 114 inches. What are the dimensions of Olivia's room in feet?

_____

Vocabulary Check **Write the vocabulary word that completes each sentence.**

37 The _____ system is a measurement system that includes units such as foot, pound, and quart.

38 To _____ means to find an equivalent measure.

39 **Writing in Math** Explain how to convert 288 inches to yards.

_____

_____

▶ Spiral Review

40 Draw a line segment that is 48 millimeters long. (Lesson 4-1, p. 142)

```
|||||||||||||||||||||||||||||||||||||||||||||||
 0    1    2    3    4    5    6
cm
```

Convert. (Lesson 4-2, p. 149)

41 980 km = _____ m

42 85.2 cm = _____ m

Measure the length of the line segment to the nearest $\frac{1}{4}$ inch.
(Lesson 4-3, p. 157)

43 The line segment is about _____ inches in length.

STOP

1 Draw a line segment that has a length of $4\frac{1}{2}$ inches.

2 Measure the length of the line segment to the nearest $\frac{1}{4}$ inch.

_____

3 Measure the length of the dragonfly to the nearest $\frac{1}{2}$ inch.

_____

Convert using a table.

4 3 mi = _____ in.

miles	1	2	3
yards			

5 15 ft = _____ yd

yards					
feet	3	6	9	12	15

Convert.

6 12 yd = _____ in.

7 2 mi = _____ yd

8 96 in. = _____ ft

9 42 ft = _____ yd

Solve.

10 **NUMBER SENSE** The community pool measures 25 yards long. How many inches long is the pool?

_____

Vocabulary and Concept Check

benchmark, *p. 149*

centimeter, *p. 142*

convert, *p. 149*

customary system, *p. 157*

foot, *p. 157*

inch, *p. 157*

kilometer, *p. 142*

length, *p. 142*

meter, *p. 142*

metric system, *p. 142*

mile, *p. 157*

millimeter, *p. 142*

yard, *p. 157*

Write the vocabulary word that completes each sentence.

1 The _____ is a metric unit of length equal to one thousand meters.

2 The _____ includes units such as foot, pound, and quart.

3 A(n) _____ is a customary unit of length equal to 12 inches.

4 A(n) _____ is an object or number used as a guide to estimate or reference.

5 The _____ is a decimal system of weights and measures.

6 To _____ is to find an equivalent measure.

7 A(n) _____ is a customary unit of length equal to 5,280 feet or 1,760 yards.

Label each diagram below by writing the word for the abbreviation.

8 _____   **9** _____

$$1 \text{ m} = 1,000 \text{ mm}$$

10 _____   **11** _____

$$1 \text{ yd} = 36 \text{ in.}$$

Lesson Review

4-1 Metric Length (pp. 142–148)

Measure the length of the object to the nearest centimeter.

12 The orange slice is _____ long.

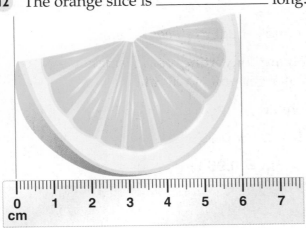

Measure the length of the object to the nearest millimeter.

13 The earthworm is

_____ long.

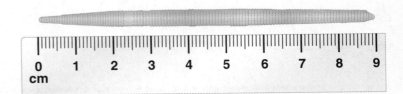

4-2 Unit Conversions: Metric Length (pp. 149–155)

14 Convert 6.5 meters to millimeters.

1000			1	0.1	0.01	0.001
thousands			ones	tenths	hundredths	thousandths
kilo (km)			meters (m) •	deci (dm)	centi (cm)	milli (mm)

6.5 m = _____ mm

Example 1

Find the length of the pencil to the nearest millimeter.

1. Line up the "zero mark" of the ruler with the left end of the pencil.

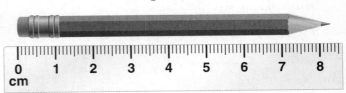

2. Read the number on the ruler that lines up with the right end.

The pencil is about 83 millimeters long.

Example 2

Convert 4 centimeters to meters.

1. Use a chart. Place 4 in the cm column.

1000			1	0.1	0.01	0.001
thousands			ones	tenths	hundredths	thousandths
kilo (km)			meters (m) 0 •	deci (dm) 0	centi (cm) 4	milli (mm)

2. Place zeros in the m and dm columns.

3. Read the number from the chart for the conversion. **4 cm = 0.04 m**

Convert.

15

$\times 1000 \quad \times 100 \quad \times 10$

larger units → km ⟲ m ⟲ cm ⟲ mm ← smaller units

$\div 1000 \quad \div 100 \quad \div 10$

3 km = _____ m

_____ m = 1 km

Multiply or divide? _____

$3 \times 1{,}000 =$ _____

3 km = _____ m

Convert.

16 4.5 mm = _____ m

17 3 m = _____ km

18 57 mm = _____ cm

19 7 m = _____ cm

4-3 Customary Length (pp. 157–162)

Measure the length of the item to the nearest $\frac{1}{4}$ inch.

20 The domino is _____ long.

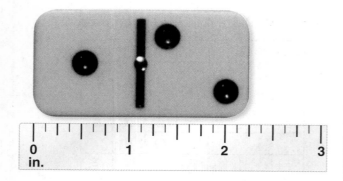

Example 3

Convert 9.5 millimeters to centimeters.

$\times 1000 \quad \times 100 \quad \times 10$

larger units → km ⟲ m ⟲ cm ⟲ mm ← smaller units

$\div 1000 \quad \div 100 \quad \div 10$

1. 1 mm is equal to 10 centimeters.

2. You are converting from a smaller unit to a larger unit, so divide.

3. Convert.

   $9.5 \div 10 = 0.95$

   9.5 mm = 0.95 cm

Example 4

Find the length of the battery to the nearest $\frac{1}{2}$ inch.

1. Line up the "zero mark" of the ruler with the left end of the battery.

2. Read the number on the ruler that lines up with the right end.

The battery is about $1\frac{1}{2}$ inches long.

4-4 Unit Conversions: Customary Length (pp. 163–168)

Convert using a table.

21 5 ft = _____ in.

feet	1	2	3	4	5
inches					

Convert.

22 6 yd = _____ ft

1 yd = _____ feet

Multiply or divide? _____

6 × 3 = _____

6 yd = _____ ft

23 7 ft = _____ in.

24 15 yd = _____ ft

Convert using a table.

25 60 in. = _____ ft

feet					
inches	12	24	36	48	60

Convert.

26 30 ft = _____ yd

1 yd = _____ ft

Multiply or divide? _____

30 ÷ 3 = _____

30 ft = _____ yd

27 24 ft = _____ yd

28 72 in. = _____ ft

Example 6

Convert 3 feet to inches.

1. You are converting from feet to inches, which is a larger unit to a smaller unit. You should multiply.

2. 1 foot is equal to 12 inches.

So, 3 feet are equal to 3 × 12, or 36 inches.

Example 7

Convert 48 inches to feet.

1. You are converting from inches to feet, which is a smaller unit to a larger unit. You should divide.

2. 12 inches are equal to 1 foot.

So, 48 inches are equal to 48 ÷ 12, or 4 feet.

Solve.

1   What is the height of the toy soldier at the right to the nearest centimeter? _____

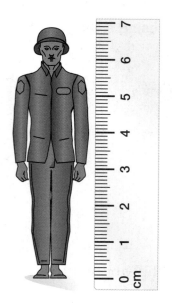

2   Draw a line segment that has a length of 59 millimeters.

|||||||||||||||||||||||||||||||||||||||||||||||||||||||
0   1   2   3   4   5   6   7   8   9   10
cm

3   Measure the length of the three buttons to the nearest $\frac{1}{2}$ inch. _____

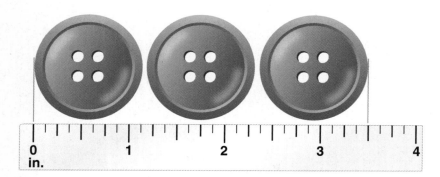

0   1   2   3   4
in.

Convert using a place-value chart or table.

4   9 cm = _____ m

1000			1	0.1	0.01	0.001
thousands			ones	tenths	hundredths	thousandths
kilo (km)			meters (m)•	deci (dm)	centi (cm)	milli (mm)

5   8.3 km = _____ m

1000			1	0.1	0.01	0.001
thousands			ones	tenths	hundredths	thousandths
kilo (km)			meters (m)•	deci (dm)	centi (cm)	milli (mm)

6   5 yd = _____ in.

yard	1	2	3	4	5
inches					

7   7 ft = _____ yd

feet	1	2	3	4	5	6	7
yards							

Convert.

8 0.28 km = _____ m

9 5 cm = _____ dm

10 3 mi = _____ ft

11 48 in. = _____ ft

Select the appropriate unit to measure the length of each object.
Write *millimeters*, *inches*, *meters*, or *miles*.

12 height of your school _____

13 width of a calculator _____

14 length of a skateboard _____

15 length of Lake Okeechobee _____

Solve.

16 **TRAVEL** It is 12 kilometers from Tayshan's house to the community swimming pool. How many meters is it to the pool?

_____

17 **SPORTS** A football field is 100 yards long. How many feet long is the football field?

_____

18 **ART** Kenisha needed strips of cloth $3\frac{3}{4}$ inches long for an art project. Draw a line segment to show the length of each piece of cloth.

Football stadium near Petaluma, California

Correct the mistakes.

19 Mr. Hopkins went to a hardware store to buy 8 yards of rope for his garden. The rope was measured in feet, so he purchased 96 feet instead. What was wrong with the purchase Mr. Hopkins made?

_____

_____

8 yards needed
× 12 feet
per yard
= 96 feet
needed

20 Show how you would correct Mr. Hopkins' mistake.

_____

_____

STOP

Choose the best answer and fill in the corresponding circle on the sheet at right.

1 Find the length of the line to the nearest millimeter.

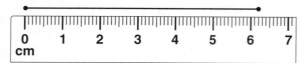

   A 6 mm   **C** 70 cm

   B 62 mm   **D** 72 mm

2 246 centimeters = _____ meters

   A 0.00246   **C** 24.6

   B 2.46   **D** 2,460

3 Find the length of the line to the nearest $\frac{1}{2}$ inch.

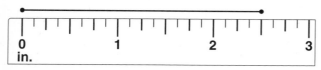

   A $1\frac{1}{2}$ in.   **C** $2\frac{1}{2}$ in.

   B $3\frac{1}{2}$ in.   **D** $4\frac{1}{2}$ in.

4 When completed, a road will be 7.36 kilometers long. What is the length in meters?

   A 0.736 m   **C** 736 m

   B 73.6 m   **D** 7,360 m

5 803 millimeters = _____ meters

   A 0.0803   **C** 8.03

   B 0.803   **D** 80.3

6 Find the length of the line to the nearest centimeter.

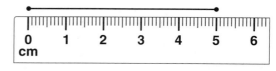

   A 3 cm   **C** 5 cm

   B 4 cm   **D** 6 cm

7 The ladder is 60 inches long. How many feet long is the ladder?

   A 3 ft

   B 5 ft

   C 6 ft

   D 7 ft

8 Javon rode his bike 2 miles to Zina's house. Together they rode another 3 miles to the park. Which sentence shows how many feet Javon traveled to get to the park?

   A $5 \times 5{,}280 = 26{,}400$ ft

   B $5 \times 12 = 60$ ft

   C $2 + 3 = 5$ ft

   D $5 \times 100 = 500$ ft

GO ON

9 Find the length of the line to the nearest $\frac{1}{4}$ inch.

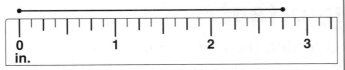

A $1\frac{1}{4}$ C 3

B $2\frac{3}{4}$ D $3\frac{1}{4}$

---

10 What customary unit of length would you use to measure a person's height?

A centimeters   C foot

B yard   D mile

---

11 Berto and his friends are enjoying the outdoor activity shown at the right. Their sled traveled 42 feet. Convert 42 feet to yards.

A 14 yd   C 17 yd

B 15 yd   D 20 yd

---

12 What metric unit of length would you use to measure the length of an ant?

A millimeter   C meter

B centimeter   D kilometer

ANSWER SHEET

Directions: Fill in the circle of each correct answer.

1   Ⓐ Ⓑ Ⓒ Ⓓ
2   Ⓐ Ⓑ Ⓒ Ⓓ
3   Ⓐ Ⓑ Ⓒ Ⓓ
4   Ⓐ Ⓑ Ⓒ Ⓓ
5   Ⓐ Ⓑ Ⓒ Ⓓ
6   Ⓐ Ⓑ Ⓒ Ⓓ
7   Ⓐ Ⓑ Ⓒ Ⓓ
8   Ⓐ Ⓑ Ⓒ Ⓓ
9   Ⓐ Ⓑ Ⓒ Ⓓ
10   Ⓐ Ⓑ Ⓒ Ⓓ
11   Ⓐ Ⓑ Ⓒ Ⓓ
12   Ⓐ Ⓑ Ⓒ Ⓓ

Success Strategy

Find key words or phrases in each question that will help you choose the correct answer. For example, pay attention to the units the question is asking you to convert.

STOP

Two-Dimensional Figures

Why is area important?

Many careers require you to find the area of a surface. Painters must measure the surface area of rooms in order to know how much paint to buy.

STEP 2 Preview

Get ready for Chapter 5. Review these skills and compare them with what you will learn in this chapter.

What You Know	What You Will Learn
You know how to measure the lengths of items. 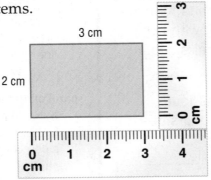	*Lessons 5-3 and 5-4* To find the area of a rectangle, multiply the length by the width. Area = length × width Area = 3 centimeters × 2 centimeters Area = 6 square centimeters
You know that you can make a parallelogram into a rectangle. Step 1: Step 2: Step 3:	*Lesson 5-5* To find the area of a parallelogram, multiply the base by the height. Area = base × height Area = 6 inches × 5 inches Area = 30 square inches 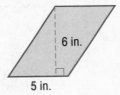
You know that you can separate a rectangle into two triangles.	*Lesson 5-6* A triangle is half of a rectangle. So the area of a triangle is one-half the area of a rectangle with the same base and height. Area = base × height Area = $\frac{1}{2}$ × base × height 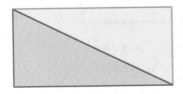 $A = \frac{1}{2} \times 5 \text{ ft} \times 4 \text{ ft} = 10 \text{ ft}^2$

Quadrilaterals

KEY Concept

Quadrilaterals have four sides and four angles. Some quadrilaterals have special names.

Type	Example	Description
rectangle		A rectangle has four right angles, with two pairs of equal sides.
square		A square has four right angles. All sides are equal.
parallelogram		The opposite sides of a parallelogram are parallel and equal in length. Opposite angles of each side are also congruent.
rhombus *These marks show equal sides.*		All four sides of a rhombus are equal. Opposite sides are parallel.
trapezoid *This symbol indicates parallel sides.*		A trapezoid has only one pair of opposite sides parallel.

VOCABULARY

congruent
line segments that have the same length or angles that have the same measure

parallel lines
lines that are the same distance apart; parallel lines do not meet or cross

parallelogram
a quadrilateral that has both pairs of opposite sides congruent and parallel

quadrilateral
a shape that has four sides and four angles

rhombus
a parallelogram with four congruent sides

trapezoid
a quadrilateral with one pair of opposite sides parallel

Quadrilaterals can be classified by the size of their angles and the length of their sides. Figures often have symbols that indicate if there are **congruent** parts.

parallel lines

congruent sides

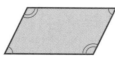
congruent angles

Example 1

Classify the figure in as many ways as possible.

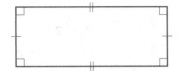

1. Look at the figure.

2. Are opposite sides equal?
   yes

3. Are any of the opposite sides parallel?
   yes

4. Does the figure have exactly one pair of parallel sides?   no

5. The figure can be classified as a parallelogram and a rectangle.

Classify the figure in as many ways as possible.

1. Look at the figure.

2. Are opposite sides equal? _____

3. Are any of the opposite sides parallel?

   _____

4. Does the figure have exactly one pair of parallel sides? _____

5. The figure can be classified as a

   _____.

Example 2

Identify the figure.

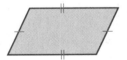

1. The figure has four sides.
   The figure is a quadrilateral.

2. The opposite sides are parallel and equal in length.
   The figure is a parallelogram or a rectangle.

3. The opposite angles of each side are congruent.
   The figure is a parallelogram.

Identify the figure.

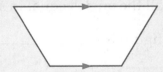

1. The figure has _____ sides.

   The figure is a(n) _____.

2. There is _____ pair of parallel sides.

3. There are _____ right angles.

   The figure is a(n) _____.

GO ON

Who is Correct?

Draw a trapezoid.

Jermaine

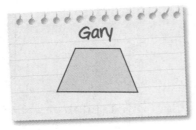
Gary

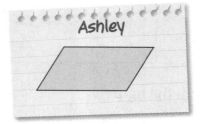
Ashley

Circle correct answer(s). Cross out incorrect answer(s).

Guided Practice

Classify each quadrilateral in as many ways as possible.

1

1. Are opposite sides equal? _____

2. Are any of the opposite sides parallel? _____

3. Are all the angles equal in size? _____

4. The figure is a _____.

2

1. Are opposite sides equal? _____

2. Are any of the opposite sides parallel? _____

3. Does the figure have exactly one pair
   of parallel sides? _____

4. The figure is a _____.

3 Identify the figure.

Step 1 The figure has _____ sides.

The figure is a(n) _____.

Step 2 Are any of the sides parallel? _____

Step 3 There are _____ pairs of parallel sides.

The figure is a(n) _____.

Identify the figure.

4 The figure has _____ sides.

The figure is a(n) _____.

The opposite _____ are parallel and equal in length.

The opposite _____ of each side are congrent.

The figure is a(n) _____.

5 The figure has _____ sides.

The figure is a(n) _____.

_____ sides are equal.

The opposite _____ are parallel.

The figure is a(n) _____.

GO ON

Step by Step Problem-Solving Practice

Solve.

Problem-Solving Strategies
- ☑ Use a diagram.
- ☐ Look for a pattern.
- ☐ Guess and check.
- ☐ Act it out.
- ☐ Work backward.

6 MUSIC Alonso plays the glockenspiel in the marching band. What quadrilateral figure describes the shape of his instrument?

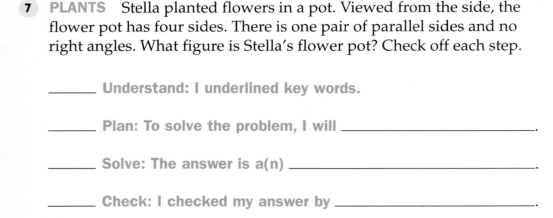

Understand   Read the problem. Write what you know. The figure of the glockenspiel is a _____.

Plan   Pick a strategy. One strategy is to use a diagram.

Solve   Trace the outline of the figure. There are _____ sides. There is _____ pair of parallel sides. The figure is a(n) _____.

Check   Review the definition of the figure you named.

7 PLANTS Stella planted flowers in a pot. Viewed from the side, the flower pot has four sides. There is one pair of parallel sides and no right angles. What figure is Stella's flower pot? Check off each step.

_____ Understand: I underlined key words.

_____ Plan: To solve the problem, I will _____.

_____ Solve: The answer is a(n) _____.

_____ Check: I checked my answer by _____.

8 Reflect The word *quadrilateral* has two parts. *Quad* means "four" and *lateral* means "side or relating to the side." In your own words, explain the meaning of the word quadrilateral.

_____

_____

Skills, Concepts, and Problem Solving

9 Circle the quadrilaterals.

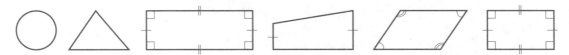

10 Circle the rectangles.

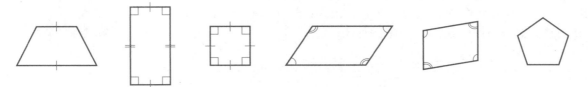

11 Circle the parallelograms.

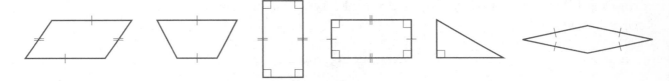

Identify each figure.

12

_____

13

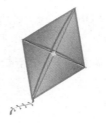

_____

14

_____

15

_____

GO ON

Identify each figure.

16

_____

17

_____

18 **SAFETY**   Alan's science teacher asked him to tape a safety sign on the wall. All four sides of the sign are equal, and the opposite sides are parallel. What is the name of this figure?

_____

19 **HOMEWORK**   Tanisha's math teacher asks her to explain the differences between parallelograms and trapezoids. Tanisha says, "A parallelogram has only one pair of opposite parallel sides. A trapezoid has two pair of opposite parallel sides." Is Tanisha correct? Explain.

_____

_____

Vocabulary Check   **Write the vocabulary word that completes each sentence.**

20 Line segments that have the same length are _____.

21 A(n) _____ has four sides and four angles.

22 A parallelogram with four congruent sides is called a(n)

_____.

23 _____ are lines that are the same distance apart; they do not meet or cross.

24 **Writing in Math**   Describe a rectangle in words.

_____

_____

_____

STOP

Copyright © Glencoe/McGraw-Hill, a division of The McGraw-Hill Companies, Inc.

186   **Chapter 5** Two-Dimensional Figures

Triangles

KEY Concept

Triangles can be named by the measure of their angles and lengths of their sides.

For example, a triangle with one **right angle** and two congruent sides is called an isosceles right triangle.

Classify Triangles by Angles

acute triangle
three angles less than 90°

obtuse triangle
one angle greater than 90°

right triangle
one 90° angle

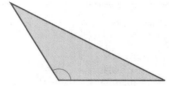
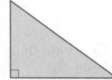

Classify Triangles by Sides

equilateral triangle
all sides are congruent

isosceles triangle
at least two sides are congruent

scalene triangle
no congruent sides

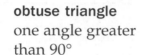
4 cm
4 cm
4 cm

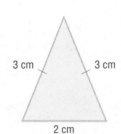
3 cm
3 cm
2 cm

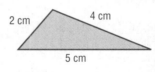
2 cm
4 cm
5 cm

VOCABULARY

acute angle
an angle with a measure greater than 0° and less than 90°

congruent
line segments that have the same length, or angles that have the same measure

obtuse angle
an angle that measures greater than 90° but less than 180°

right angle
an angle with the measure of 90°

GO ON

Example 1

Name the triangle by its sides.

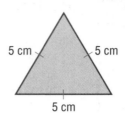

1. The triangle has **3** congruent sides.

2. The triangle is an **equilateral triangle**.

YOUR TURN!

Name the triangle by its sides.

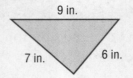

1. The triangle has _____ congruent sides.

2. The triangle is a(n) _____.

Example 2

Name the triangle by its angles.

1. Use a right angle to show 90°. Compare the right angle to the angles of the triangle.

2. There are two angles less than 90°.

3. There is one angle greater than 90°.

4. There are no right angles.

5. The triangle is an obtuse triangle.

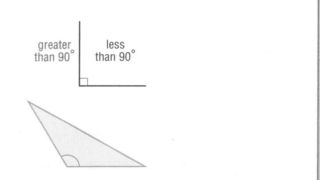

YOUR TURN!

Name the triangle by its angles.

1. Use a right angle to show 90°. Compare the right angle to the angles of the triangle.

2. There are _____ angles less than 90°.

3. There are _____ angles greater than 90°.

4. There are _____ right angles.

5. The triangle is a(n) _____ triangle.

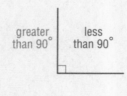

Who is Correct?

Draw an acute triangle.

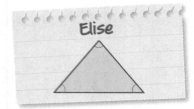

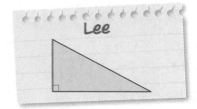

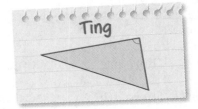

Circle correct answer(s). Cross out incorrect answer(s).

▶ Guided Practice

Name each triangle by its sides.

1   The triangle has _____ congruent sides.

The triangle is a(n) _____.

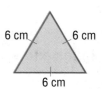

2

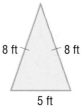

3

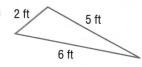

_____

_____

Name the triangle by its angles.

4   There are _____ angles less than 90°.

There are _____ angles greater than 90°.

There are _____ right angles.

The triangle is a(n) _____.

5

6

_____

_____

GO ON

7 Name the triangle by its angles and sides.

Step 1 There are _____ congruent sides.

Step 2 There are _____ acute angles.

Step 3 There are _____ right angles.

Step 4 There is _____ obtuse angle.

Step 5 The triangle is a(n) _____ triangle.

Name each triangle by its angles and sides.

8 number of congruent sides _____

number of acute angles _____

number of right angles _____

number of obtuse angles _____

The triangle is a(n) _____ triangle.

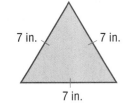

9

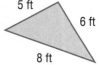

_____

10

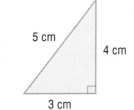

_____

Step by Step Problem-Solving Practice

Solve.

11 **HOME ARTS** Rondell made an apple turnover in Home Arts that looked like the photo at the right. He thought it looked like a triangle so he traced it. Classify the triangle he traced by its angles.

Problem-Solving Strategies
- ☑ Use a diagram.
- ☐ Look for a pattern.
- ☐ Guess and check.
- ☐ Act it out.
- ☐ Work backward.

Understand	Read the problem. Write what you know.
	Rondell made _____.
Plan	Pick a strategy. One strategy is to use a diagram.
Solve	There are _____ angles less than 90°.
	There are _____ angles greater than 90°.
	There is _____ right angle.
	The figure is a(n) _____.
Check	Review the definition of the figure you named.

12 **BASEBALL** Desiree bought a pennant banner at a baseball game. If the triangle is classified by its angles, what type of triangle is shown? Check off each step.

_____ Understand: I underlined key words.

_____ Plan: To solve the problem, I will _____.

_____ Solve: The answer is _____.

_____ Check: I checked my answer by _____.

13 **Reflect** Can a triangle be both isosceles and obtuse? Explain.

_____

_____

GO ON

Name each triangle by its sides.

14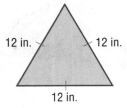

3 cm
6 cm
7 cm

_____

15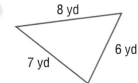

8 yd
6 yd
7 yd

_____

16

12 in.   12 in.

12 in.

_____

17

7 ft   5 ft

5 ft

_____

Name each triangle by its angles.

18

_____

19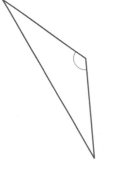

_____

20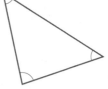

_____

21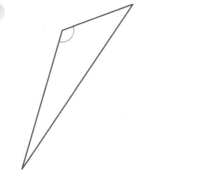

_____

22 **ART** Sandy's art teacher asked her class to cut out a triangle with sides that measure 9 inches, 7 inches, and 12 inches. What type of triangle did the students cut out? _____

23 Name the triangle by its angles and sides.

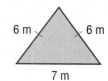

_____

24 **MUSIC** The musical instrument pictured on the right is in the shape of a triangle. All three angles of the triangle are less than 90°. Name the type of triangle pictured.

_____

Vocabulary Check **Write the vocabulary word that completes each sentence.**

25 Angles that have the same measure are _____.

26 A triangle with an angle that measures greater than 90° but less than 180° is a(n) _____.

27 A(n) _____ has three equal sides.

28 **Writing in Math** Can a triangle be both a right triangle and an acute triangle?

_____

_____

_____

 Spiral Review

Identify each figure. (Lesson 5-1, p. 180)

29

_____

30

_____

STOP

Write the name of each figure.

1

_____

2

_____

3

_____

4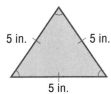

_____

Name each triangle by its sides or angles.

5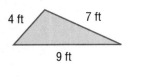
4 ft   7 ft
9 ft

_____

6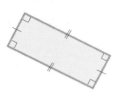
5 in.   5 in.
5 in.

_____

7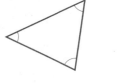

_____

8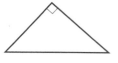

_____

Solve.

9 **FIGURES** Mindy bought a purse for her friend's birthday. What type of quadrilateral is the red part of the purse?

_____

10 Can a right triangle also be equilateral? Explain.

_____

_____

_____

Introduction to Area

KEY Concept

The **area** of a figure is the number of **square units** needed to cover a surface.

To find the area of a figure, you can count the number of square units the figure covers.

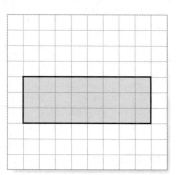

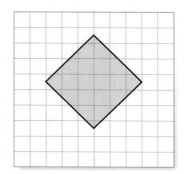

The area of the rectangle is 24 square units.

The area of the figure is about 18 square units.

VOCABULARY

area
the number of square units needed to cover the surface enclosed by a geometric figure

square unit
a unit for measuring area

The units of area are square units.

Example 1

Find the area of the rectangle.

1. Count the number of squares the rectangle covers.

2. The area of the rectangle is 30 square units.

3. Check your answer.
   Count the squares in the top row. 6
   Count the number of rows. 5

4. Add the rows to find the area.

   6 + 6 + 6 + 6 + 6 = 30

   The area of the figure is about 30 square units.

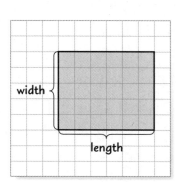

GO ON

YOUR TURN!

Find the area of the rectangle.

1. Count the number of squares the rectangle covers.

2. The area of the rectangle is _____ square units.

3. Check your answer. Count the squares in the top row. _____

   Count the number of rows. _____

4. Add the rows to find the area.

   _____ + _____ + _____ + _____ + _____ + _____ + _____ + _____ + _____ = _____

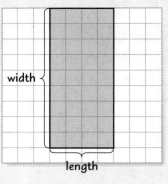

width

length

Example 2

Estimate the area of the figure.

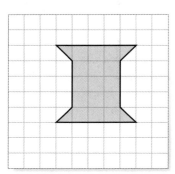

1. Count the number of whole squares the figure covers.

   The figure covers 15 whole squares.

2. Count the number of half squares the figure covers.

   The figure covers 4 half squares.

   4 half squares = 2 whole square(s)

3. Add the number of whole squares.
   15 + 2 = 17

The area of the figure is about 17 square units.

YOUR TURN!

Estimate the area of the figure.

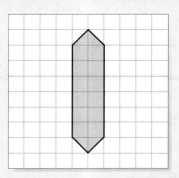

1. Count the number of whole squares the figure covers.

   The figure covers _____ whole squares.

2. Count the number of half squares the figure covers.

   The figure covers 4 _____ square(s).

   _____ half squares = _____ whole square(s)

3. Add the number of whole squares.

   _____ + _____ = _____

The area of the figure is about _____ square units.

Who is Correct?

Find the area of the square.

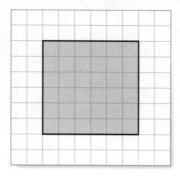

Lorenzo	Troy	Bianca
6 + 6 + 6 + 6 + 6 = 30	6 + 6 + 6 + 6 + 6 + 6 = 36	6 + 6 + 6 + 6 = 24
30 square units	36 square units	24 square units

Circle correct answer(s). Cross out incorrect answer(s).

▶ Guided Practice

Draw a figure that has the given area.

1   21 square units

2   49 square units

Step by Step Practice

3   Estimate the area of the figure.

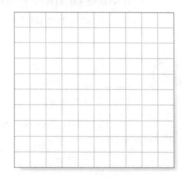

Step 1   Count the number of whole squares the figure covers. The figure covers _____ whole squares.

Step 2   Count the number of half squares.
_____ half squares = _____ whole square(s)

Step 3   Add the number of whole squares.
_____ + _____ = _____

The area of the figure is _____ square units.

GO ON

Find the area of each figure.

4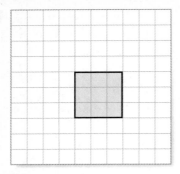

The area of the square is
_____ square units.

5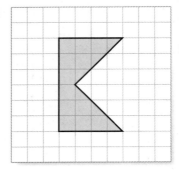

The area of the figure is about
_____ square units.

6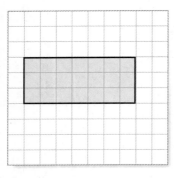

The area of the rectangle is
_____ square units.

7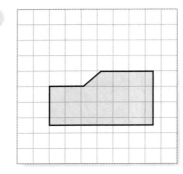

The area of the figure is about
_____ square units.

Step by Step Problem-Solving Practice

Problem-Solving Strategies
- ☑ Draw a diagram.
- ☐ Look for a pattern.
- ☐ Guess and check.
- ☐ Act it out.
- ☐ Solve a simpler problem.

Solve.

8 GEOMETRY What is the area of a rectangle that has sides
of 8 units and 4 units?

Understand Read the problem. Write what you know.
A rectangle has sides of _____ units and
_____ units.

Plan Pick a strategy. One strategy is to draw
a diagram. Draw a rectangle that has
sides of 8 units and 4 units.

Solve Count the number of squares the figure covers.
The area of the rectangle is _____ square units.

Check You can multiply 8 and 4 because there are
4 rows of 8 units. $8 \times 4 =$ _____

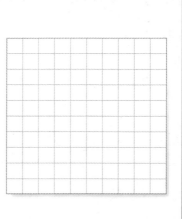

9 **PATIOS**   Talia's patio in her backyard measures 9 feet by 7 feet. What is the area of the patio in Talia's backyard? Check off each step.

_____ Understand: I underlined the keywords.

_____ Plan: To solve the problem I will _____.

_____ Solve: The answer is _____.

_____ Check: I checked my answer by _____.

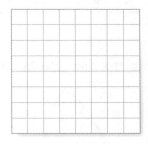

10 **ART**   Selina is using one-inch square clay tiles to create a plate in ceramics class. She wants her plate to have 8 rows of 8 tiles. What is the area of her plate? _____

11 **Reflect**   Is the area of the figure at the right 23 square units? Explain.

_____

_____

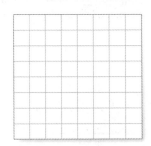

▶ Skills, Concepts, and Problem Solving

Draw a figure that has the given area.

12 16 square units

13 24 square units

Find the area of each figure.

14 _____ square units

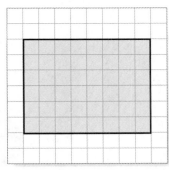

15 _____ square units

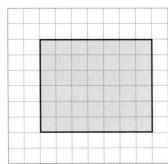

GO ON

Solve.

16 **BASEBALL** The batter's box in baseball measures about 6 feet by 4 feet. What is the area of the batter's box? _____

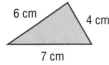

Vocabulary Check **Write the vocabulary word that completes each sentence.**

17 A(n) _____ is a unit for measuring area.

18 _____ is the number of square units needed to cover the surface enclosed by a geometric figure.

19 **Writing in Math** Explain how to find the length and width of a rectangle with an area of 36 square units.

_____

_____

_____

▶ Spiral Review

Identify each figure. (Lesson 5-1, p. 180)

20

_____

21

_____

Name the triangle by its sides. (Lesson 5-2, p. 187)

22

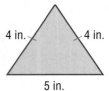

4 in. 4 in.

5 in.

_____

23

6 cm   4 cm

7 cm

_____

Solve.

24 **SCHOOL BAND** The high school marching band made a triangular formation during their routine. All the angles they formed were less than 90°. What type of triangle did they form? _____

Area of a Rectangle

KEY Concept

Find the **area** of a rectangle using the formula below.

ℓ is the length of
the rectangle.

A is the area of
the rectangle. ⟶ $A = \ell \times w$ ⟵ w is the width of the
rectangle.

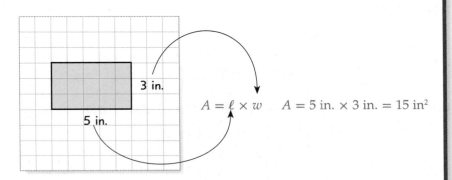

$A = \ell \times w$ $A = 5 \text{ in.} \times 3 \text{ in.} = 15 \text{ in}^2$

3 in.

5 in.

The area of the rectangle is 15 square inches.

The units of area are **square units**.

Example 1

What is the area of the rectangle?

1. The length of the rectangle is 6 centimeters,
   and the width is 4 centimeters.

2. Substitute these values into the formula. Multiply.

   $A = \ell \times w$
   $A = 6 \text{ cm} \times 4 \text{ cm}$
   $A = 24 \text{ cm}^2$

The area of the rectangle is 24 square centimeters.

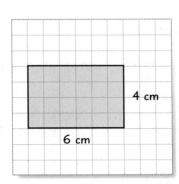

4 cm

6 cm

GO ON

YOUR TURN!

What is the area of the rectangle?

1. The length of the rectangle is _____ feet, and the width is _____ feet.

2. Substitute these values into the formula. Multiply.

   $A = \ell \times w$

   $A =$ _____ feet × _____ feet

   $A =$ _____ feet²

The area of the rectangle is _____ square feet.

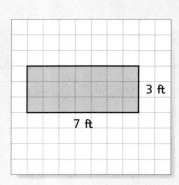

3 ft

7 ft

Example 2

What is the area of the square?

1. The length of the square is 5 km, and the width is 5 km.

2. Substitute these values into the formula. Multiply.

   $A = \ell \times w$
   $A = 5 \text{ km} \times 5 \text{ km}$
   $A = 25 \text{ km}^2$

The area of the square is 25 square km.

5 km

5 km

YOUR TURN!

What is the area of the square?

1. The length of the square is _____ feet, and the width is _____ feet.

2. Substitute these values into the formula. Multiply.

   $A = \ell \times w$
   $A =$ _____ ft × _____ ft
   $A =$ _____ ft²

The area of the square is _____ square feet.

4 ft

4 ft

Who is Correct?

What is the area of the rectangle at the right?

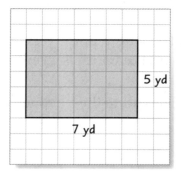

5 yd

7 yd

Juan
$A = 7 \times 5 = 35 \text{ yd}^2$

Jarred
$A = 7 + 5 = 12 \text{ yd}^2$

Fidel
$A = 7 \times 7 = 49 \text{ yd}^2$

Circle correct answer(s). Cross out incorrect answer(s).

▶ Guided Practice

Draw a rectangle for each given area.

1   12 in²

2   32 m²

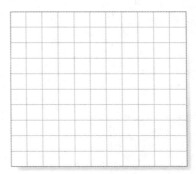

Step (by) Step *Practice*

3   What is the area of the square?

> **Step 1**   The length of the square is _____ centimeters, and the width _____ centimeters.
>
> **Step 2**   Substitute these values into the formula. Multiply.
>
> $A = \ell \times w$
>
> $A =$ _____ cm $\times$ _____ cm
>
> $A =$ _____ cm²
>
> The area of the rectangle is _____ square centimeters.

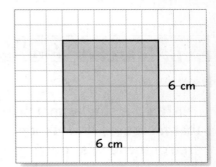

6 cm

6 cm

Find the area of each rectangle.

4   The length is _____ meters. The width is _____ meters.

$A =$ _____ m $\times$ _____ m

$A =$ _____ m²

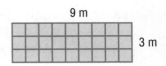

9 m

3 m

5   The length of the rectangle is _____ inches. The width is _____ inches.

$A = \ell \times w$

$A =$ _____ in. $\times$ _____ in.

$A =$ _____ in²

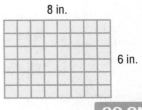

8 in.

6 in.

GO ON

Find the area of each figure.

6 $A =$ _____

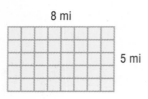
8 mi

5 mi

7 $A =$ _____

10 cm

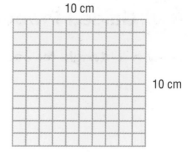

10 cm

8 $A =$ _____

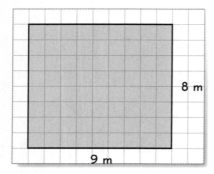

8 m

9 m

9 $A =$ _____

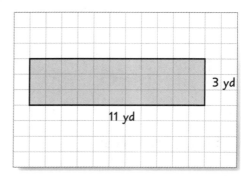

3 yd

11 yd

Step by Step Problem-Solving Practice

Solve.

10 **TENNIS** A singles match in tennis is played on a court that measures 78 feet long and 27 feet wide. What is the area of the tennis court?

Understand Read the problem. Write what you know. The length of the tennis court is _____ feet, and the width of the court is _____ feet.

Plan Pick a strategy. One strategy is to use a formula. Substitute values for length and width into the area formula.

Solve Use the formula.

$A = \ell \times w$

$A =$ _____ ft $\times$ _____ ft

$A =$ _____ ft²

The area of a tennis court is _____ square feet.

Check Use a calculator to check your answer.

Problem-Solving Strategies

☑ Use a formula.
☐ Look for a pattern.
☐ Guess and check.
☐ Act it out.
☐ Solve a simpler problem.

11   NURSING HOME   Ms. Guzman's art class made quilts for the community nursing home. Each quilt was 60 inches long and 80 inches wide. What is the area of each quilt the class made for the nursing home?

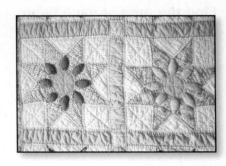

Check off each step.

_____ Understand: I underlined the key words.

_____ Plan: To solve the problem I will _____ .

_____ Solve: The answer is _____ .

_____ Check: I checked my answer by _____

_____ .

12   BAKING   Randall made garlic bread for dinner. He put the garlic bread on a baking sheet that measured 9 inches wide by 13 inches long. What was the area of the baking sheet? _____

13   Reflect   Can two rectangles have the same area but different lengths and widths? Explain.

_____

_____

_____

▶ **Skills, Concepts, and Problem Solving**

Draw a rectangle that has the given area.

14 16 square units

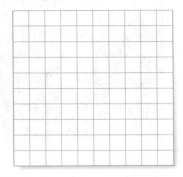

15 30 square units

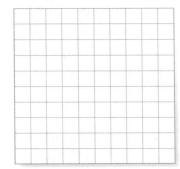

GO ON

Find the area of each figure.

16 A = _____

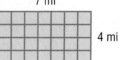
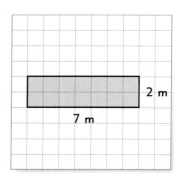
7 mi

4 mi

17 A = _____

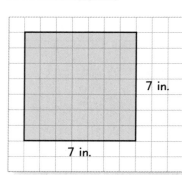
6 yd

3 yd

18 A = _____

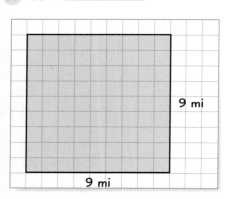
2 m

7 m

19 A = _____

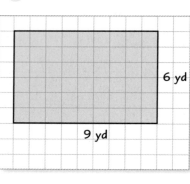
7 in.

7 in.

20 A = _____

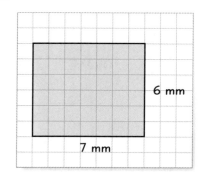
6 mm

7 mm

21 A = _____

9 mi

9 mi

22 A = _____

6 yd

9 yd

23 A = _____

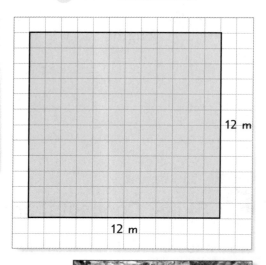
12 m

12 m

Solve.

24 **VOLUNTEERS**   Ed received a certificate for his community volunteer work from the mayor of his city. The certificate was 28 cm long and 22 cm wide. What is the area of Ed's certificate?

_____

25 **GARDENING**   Marcie made a vegetable garden in her backyard shown at the right. What is the area of Marcie's garden?

_____

GARDENING   The vegetable garden is 30 feet long and 27 feet wide.

Vocabulary Check **Write the vocabulary word that completes each sentence.**

26 A(n) _____ has opposite sides that are equal and parallel. It is a quadrilateral with four right angles.

27 _____ is the number of square units needed to cover the surface enclosed by a geometric figure.

28 A(n) _____ is a rectangle with four congruent sides.

29 **Writing in Math** Explain how to find the area of a rectangle.

_____

_____

_____

 Spiral Review

Solve.

30 **CONSTRUCTION** Jen and her friends built a tree house in her backyard. The floor of the tree house was 12 feet long by 8 feet wide. What was the area of the tree house's floor?
(Lesson 5-3, p. 195)

_____

Identify the figure. (Lesson 5-1, p. 180)

31
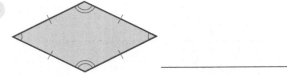
_____

Name the triangle by its angles. (Lesson 5-2, p. 187)

32
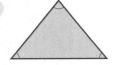
_____

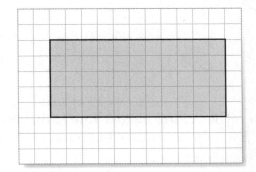

Find the area of the figure. (Lesson 5-3, p. 195)

33 $A =$ _____ square units

Draw a figure that has the given area.

1 54 square units

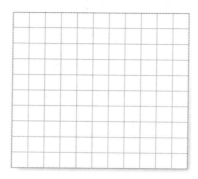

2 42 square units

Find the area of each figure.

3

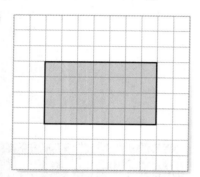

The area of the rectangle is
_____ square units.

4

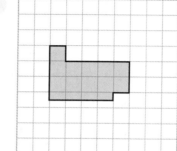

The area of the figure is
_____ square units.

Find the area of each rectangle.

5 $A =$ _____ m²

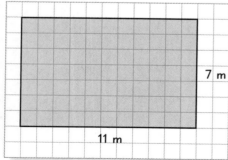

7 m

11 m

6 $A =$ _____ yd²

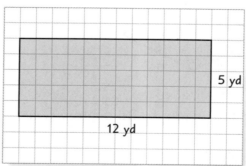

5 yd

12 yd

Solve.

7 **ROOMS** The ceiling tiles in Toni's kitchen are the shape of a
square. Each tile has sides that measure 54 centimeters. What is the
area of each tile? _____

Area of a Parallelogram

KEY Concept

Copyright © Glencoe/McGraw-Hill, a division of The McGraw-Hill Companies, Inc.

parallelogram

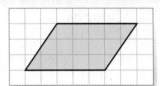

In a parallelogram, *b* represents the base, and *h* represents the height. Cut a triangle from the parallelogram along the dashed line. Place the triangle on the other side, next to the right edge of the parallelogram.

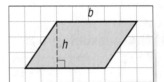 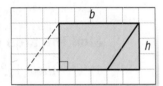

Notice that the new shape is a rectangle. So, the formulas for the areas of parallelograms and rectangles are similar.

A is the area of the parallelogram.

This is like the area of a rectangle, except the length is *b* and width is *h*.

$$A = \ell \times w$$
$$A = b \times h$$

b is the length of the base.

h is the height.

VOCABULARY

area
  the number of square units needed to cover the surface enclosed by a geometric figure

parallelogram
  a quadrilateral that has both pairs of opposite sides congruent and parallel

square unit
  a unit for measuring area

Example 1

Find the area of the parallelogram.

1. The base is 7 feet. The height is 6 feet.

2. Substitute these values into the formula. Multiply.

   $A = b \times h$
   $A = 7 \text{ ft} \times 6 \text{ ft}$
   $A = 42 \text{ ft}^2$

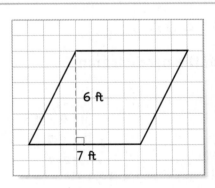

The area of the parallelogram is 42 square feet.

GO ON

Find the area of the parallelogram.

1. The base is _____ inches. The height is _____ inches.

2. Substitute these values into the formula. Multiply.

$A = b \times h$

$A = $ _____ in. $\times$ _____ in.

$A = $ _____ in²

The area of the parallelogram is _____ square inches.

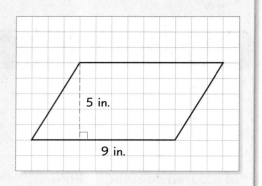
5 in.
9 in.

Example 2

Find the area of the parallelogram.

1. The base is 8 meters.
   The height is 8 meters.

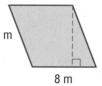
8 m
8 m

2. Substitute these values into the formula. Multiply.

$A = b \times h$

$A = 8 \text{ m} \times 8 \text{ m}$

$A = 64 \text{ m}^2$

The area of the parallelogram is 64 square centimeters.

Find the area of the parallelogram.

1. The base is

   _____ centimeters.
   The height is

   _____ centimeters.

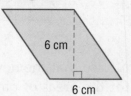
6 cm
6 cm

2. Substitute these values into the formula. Multiply.

$A = b \times h$

$A = $ _____ cm $\times$ _____ cm

$A = $ _____ cm²

The area of the parallelogram is _____ square centimeters.

Who is Correct?

Find the area of the parallelogram.

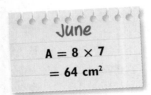
June
A = 8 × 7
= 64 cm²

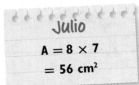
Julio
A = 8 × 7
= 56 cm²

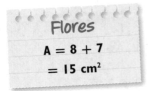
Flores
A = 8 + 7
= 15 cm²

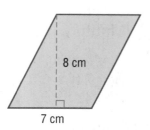
8 cm
7 cm

Circle correct answer(s). Cross out incorrect answer(s).

Guided Practice

Draw a parallelogram that has the given area.

1 36 mm²

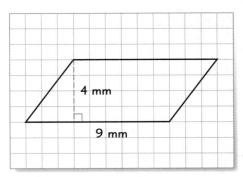

4 mm
9 mm

2 80 yd²

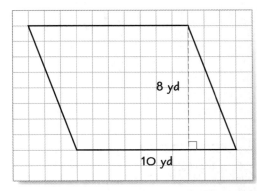

8 yd
10 yd

Step by Step Practice

Find the area of the parallelogram.

3 **Step 1** The base is _____ centimeters.
The height is _____ centimeters.

Step 2 Substitute these values into the formula. Multiply.

$A = b \times h$
$A =$ _____ cm $\times$ _____ cm
$A =$ _____ cm²

The area of the parallelogram is

_____ square centimeters.

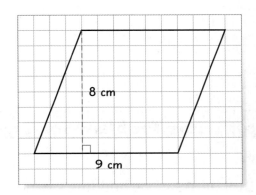

8 cm
9 cm

Find the area of each parallelogram.

4 The base is _____.
The height is _____.

$A = b \times h$
$A =$ _____ $\times$ _____
$A =$ _____

The area of the parallelogram is _____.

5 km
8 km

5 The base is _____.
The height is _____.

$A =$ _____ $\times$ _____
$A =$ _____

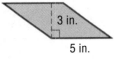

3 in.
5 in.

GO ON

6

5 mi

11 mi

$A =$ _____ × _____

$A =$ _____

7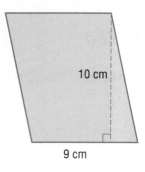

10 cm

9 cm

$A =$ _____ × _____

$A =$ _____

8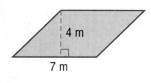

4 m

7 m

$A =$ _____

9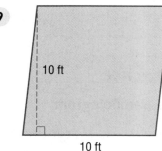

10 ft

10 ft

$A =$ _____

Step by Step Problem-Solving Practice

Solve.

10 BUILDINGS The building's windows pictured on the right are parallelograms. The windows are 48 inches wide at the base and 60 inches tall. What is the area of the windows?

Understand Read the problem. Write what you know.

The base of the window is _____ inches.

The height of the window is _____ inches.

Plan Pick a strategy. One strategy is to use a formula. Substitute the values for base and height into the area formula.

Solve Use the formula. $A = b \times h$

$A =$ _____ × _____

$A =$ _____

The area of the window is _____.

Check Use division or a calculator to check your multiplication.

Problem-Solving Strategies

☐ Draw a diagram.

☑ Use a formula.

☐ Guess and check.

☐ Solve a simpler problem.

☐ Work backward.

11 PETS The roof of Wesley's doghouse is shaped like a parallelogram. The roof has a base of 9 feet and a height of 7 feet. What is the area of the roof of the doghouse? Check off each step.

_____ **Understand: I underlined key words.**

_____ **Plan: To solve the problem, I will** _____.

_____ **Solve: The answer is** _____.

_____ **Check: I checked my answer by** _____.

12 WATCHES Rita received a watch as a present for her birthday. The face of the watch is shaped like a parallelogram with a base of 30 mm and a height of 40 mm. What is the area of the face of Rita's watch?

_____

13 Reflect Compare the area of the parallelogram to the area of the rectangle at the right. Explain.

_____

_____

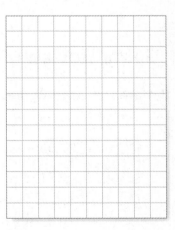

6 in.

8 in.

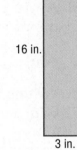

16 in.

3 in.

▶ **Skills, Concepts, and Problem Solving**

Draw a parallelogram that has the given area.

14 27 in²

15 40 mm²

Find the area of each parallelogram.

16 $A =$ _____

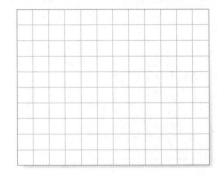

7 ft

17 ft

17 $A =$ _____

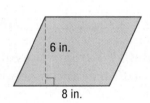

6 yd

18 yd

GO ON

Solve.

18 **STUDENT COUNCIL** Alfonso made posters in the shape of a parallelogram for the student council election. The posters were 11 inches long and 9 inches tall. What is the area of each poster?

_____

19 **HATS** The top of Mateo's graduation hat is the shape of a parallelogram. The cap has a height of 18 centimeters and a base of 23 centimeters. What is the area of the cap?

_____

Vocabulary Check **Write the vocabulary word that completes each sentence.**

20 A(n) _____ is a quadrilateral that has both pairs of opposite sides congruent and parallel.

21 A(n) _____ is a unit for measuring area.

22 **Writing in Math** Explain how to find the area of a parallelogram with a base of 12 feet and a height of 7 feet.

_____

_____

 Spiral Review

Find the area of each figure. (Lesson 5-3, p. 193)

23 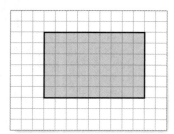 _____ **24** 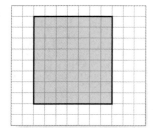 _____

Solve. (Lesson 5-4, p. 201)

25 **BASKETBALL** The high-school basketball court needed to be replaced. It is 94 feet long and 50 feet wide. What is the area of the court?

_____

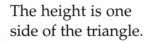

Lesson 5-6   Area of a Triangle

KEY Concept

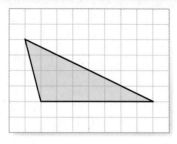

triangle

To find the area of a triangle, use what you know about the area of a **parallelogram**. You can cut a parallelogram to create two **triangles**.

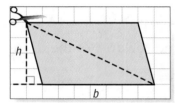 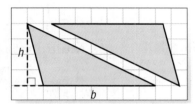

You know each triangle is $\frac{1}{2}$ the size of the parallelogram. You know the formula for the area of a parallelogram is $A = b \times h$. So, you can multiply the area of a parallelogram by $\frac{1}{2}$ to find the area of a triangle.

h is the height.

$$A = \frac{1}{2} \times b \times h$$

A is the area of the triangle.

b is the length of the base.

The location of the height of a triangle can vary. There are three possibilities.

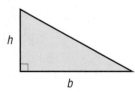 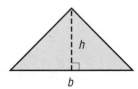 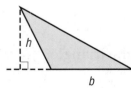

The height is one side of the triangle.

The height is inside the triangle.

The height is outside the triangle.

GO ON

Example 1

Find the area of the triangle.

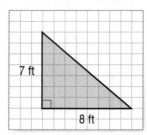

1. The base is 8 feet.

2. The height is 7 feet.

3. Substitute these values into the formula.

   $A = \frac{1}{2} \times b \times h$

   $A = \frac{1}{2} \times 8 \text{ ft} \times 7 \text{ ft}$

4. Multiply to find the area of the triangle.

   $A = 28 \text{ ft}^2$

   The area of the triangle is 28 square feet.

YOUR TURN!

Find the area of the triangle.

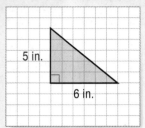

1. The base is _____ inches.

2. The height is _____ inches.

3. Substitute these values into the formula.

   $A = \frac{1}{2} \times b \times h$

   $A = \frac{1}{2} \times$ _____ in. $\times$ _____ in.

4. Multiply to find the area of the triangle.

   $A =$ _____ in^2

   The area of the triangle is

   _____ square inches.

Example 2

Find the area of the triangle.

1. The base is 3 units long.

2. The height is 4 units long.

3. Substitute these values into the formula.

   $A = \frac{1}{2} \times b \times h$

   $A = \frac{1}{2} \times 3 \times 4$

4. Multiply to find the area of the triangle.

   $A = 6 \text{ units}^2$

   The area of the triangle is 6 square units.

YOUR TURN!

Find the area of the triangle.

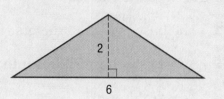

1. The base is _____ units long.

2. The height is _____ units long.

3. Substitute these values into the formula.

$$A = \frac{1}{2} \times b \times h$$

$$A = \frac{1}{2} \times \text{_____} \times \text{_____}$$

4. Multiply to find the area of the triangle.

$$A = \text{_____}$$

The area of the triangle is _____ square units.

Who is Correct?

Find the area of the triangle.

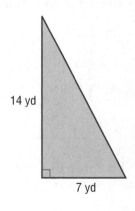

14 yd

7 yd

Tia

$A = 7 \times 14$

$= 98 \text{ yd}^2$

Manuel

$A = \frac{1}{2}(7 + 14)$

$= 10.5 \text{ yd}^2$

Luke

$A = \frac{1}{2} \times 7 \times 14$

$= 49 \text{ yd}^2$

Circle correct answer(s). Cross out incorrect answer(s).

▶ Guided Practice

Draw a triangle that has the given area.

1   12 units²

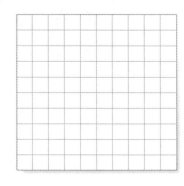

2   28 units²

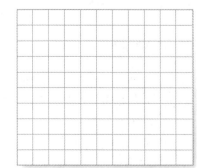

GO ON

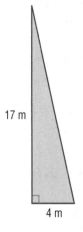

Step by Step Practice

3 Find the area of the triangle.

Step 1 The base is _____ meters. The height is _____ meters.

Step 2 Substitute these values into the formula.

$$A = \frac{1}{2} \times b \times h$$

$$A = \frac{1}{2} \times \underline{\hspace{1cm}} \text{ m} \times \underline{\hspace{1cm}} \text{ m}$$

Step 3 Multiply to find the area of the triangle.

$$A = \underline{\hspace{1cm}} \text{ m}^2$$

The area of the triangle is _____ square meters.

17 m

4 m

Find the area of each triangle.

4 The base is _____. The height is _____.

$$A = \frac{1}{2} \times b \times h$$

$$A = \frac{1}{2} \times \underline{\hspace{1cm}} \times \underline{\hspace{1cm}}$$

$$A = \underline{\hspace{1cm}}$$

The area of the triangle is _____.

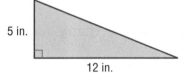

5 in.

12 in.

5 The base is _____. The height is _____.

$$A = \frac{1}{2} \times b \times h$$

$$A = \frac{1}{2} \times \underline{\hspace{1cm}} \times \underline{\hspace{1cm}}$$

$$A = \underline{\hspace{1cm}}$$

The area of the triangle is _____.

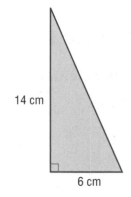

14 cm

6 cm

6 The area of the triangle is

_____.

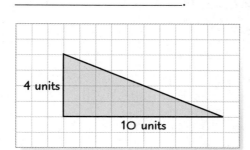

7 The area of the triangle is

_____.

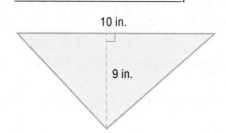

Step (by) Step Problem-Solving Practice

Solve.

8 **FLAGS** Mirna raises the flag every morning at summer camp. The flag is the shape of a triangle. It is **48** inches long at its base and has a height of 72 inches. What is the area of the camp flag?

Problem-Solving Strategies

☐ Draw a diagram.
☐ Look for a pattern.
☐ Guess and check.
☑ Use a formula.
☐ Solve a simpler problem.

Understand Read the problem. Write what you know.

The base of the flag is _____ inches.
The height of the flag is _____ inches.

Plan Pick a strategy. One strategy is to use a formula.

Substitute these values into the formula.

Solve Use the formula.

$A = \dfrac{1}{2} \times b \times h$

$A = \dfrac{1}{2} \times$ _____ $\times$ _____

$A =$ _____

The area of the flag is _____.

Check Use a calculator to check your answer.

GO ON

9   ART   The art teacher asked the students to cut out triangles for their 3-D collage. Each triangle had to be 62 millimeters long at its base and have a height of 47 millimeters. What is the area of each triangle? Check off each step.

_____ Understand: I underlined key words.

_____ Plan: To solve the problem, I will _____.

_____ Solve: The answer is _____.

_____ Check: I checked my answer by _____.

10   Belinda bought a key chain in the shape of a triangle. The key chain is 40 millimeters tall and has a base of 55 millimeters. What is the area of the key chain?

_____

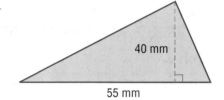

40 mm

55 mm

11   **Reflect**   Compare the area of the triangle to the area of the parallelogram at the right.

_____

_____

_____

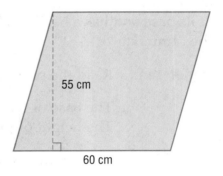

55 cm

60 cm

 Skills, Concepts, and Problem Solving

Draw a triangle that has the given area.

12   9 units²

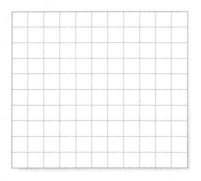

13   20 units²

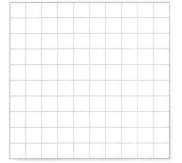

Find the area of each triangle.

14

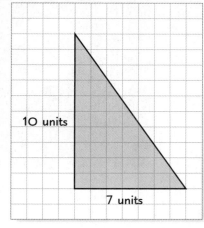

10 units

7 units

A = _____

15

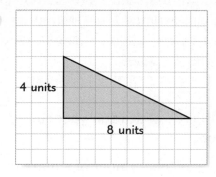

4 units

8 units

A = _____

16

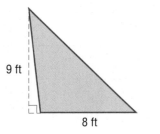

9 ft

8 ft

A = _____

17

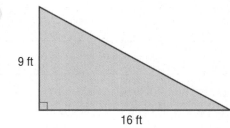

9 ft

16 ft

A = _____

18

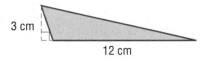

3 cm

12 cm

A = _____

19

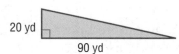

20 yd

90 yd

A = _____

Solve.

20 **WEATHER** Pearl watched a video recording of a weather report in science class. The meterologist reported on a region that had a thunderstorm watch. The region was shaped like a triangle. It measured 34 miles at the base and a height of 41 miles. What was the area of the region with the thunderstorm watch?

_____

GO ON

21 RACE Fillmore Junior High School had a relay race on field day. The race was arranged in a triangular formation. It had a base of 25 yards and a height of 50 yards. What was the area of the race?

_____

Vocabulary Check **Write the vocabulary word that completes each sentence.**

22 _____ is the number of square units needed to cover the surface enclosed by a geometric figure.

23 A(n) _____ is a polygon with three sides and three angles.

24 Writing in Math Explain how the area of a triangle is related to the area of a rectangle.

_____

_____

_____

▶ Spiral Review

Solve.

25 PAINTING Paige painted parallelograms on her two-dimensional painting. The parallelograms were 46 millimeters wide at the base and 63 millimeters tall. What was the area of each parallelogram?
(Lesson 5-5, p. 209)

_____

26 TILES Sean is placing new ceramic tiles on his bathroom floor. The ceramic tiles have four sides and four right angles. There are two sets of lines equal in length and the opposite sides are parallel. What is the name of the figure that describes the ceramic tiles?
(Lessons 5-1, p. 180)

_____

STOP

Circles

KEY Concept

Copyright © Glencoe/McGraw-Hill, a division of The McGraw-Hill Companies, Inc.

A **circle** is the set of all points in a plane that are the same distance from a point called the center.

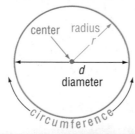

center radius
r
d
diameter
circumference

Measurement	Description	Formula
diameter and radius	The length of the **diameter** is twice the length of the **radius** of a circle.	$d = 2r$ or $r = \dfrac{d}{2}$ ↑ diameter ↑ radius
circumference	The **circumference** of a circle is equal to π times its diameter or π times twice its radius.	circumference ↙ ↘ $C = \pi d$ or $C = 2\pi r$ ↗ ↗ π is approximately **3.14**
area	The area of any circle is always equal to π times the radius squared.	$A = \pi r^2$ ↗ ↖ Area radius squared

Pi (π) is a Greek letter used to represent the ratio of the circumference to the diameter. It is not possible to write the exact value of π. To calculate using π, either use a calculator with a π button, or substitute an approximate value of π, such as 3.14.

VOCABULARY

circle
the set of all points in a plane that are the same distance from a given point called the center

circumference
the distance around a circle

diameter
the distance across a circle through its center

pi (π)
the ratio of the circumference of a circle to the diameter of the same circle; the value of π is approximately 3.14

radius
the distance from the center of a circle to any point on the circle

When you make a substitution for π, you have to change the = symbol to ≈ symbol to show that you are making an estimate.

GO ON

Example 1

Find the radius of the circle.

d = 18 cm

1. The radius is $\frac{1}{2}$ as long as the diameter.

2. One-half of 18 cm is 9 cm.

   18 cm ÷ 2 = 9 cm

3. The radius of the circle is 9 cm.

YOUR TURN!

Find the radius of the circle.

d = 46 ft

1. The radius is _____ as long as the diameter.

2. One-half of _____ is _____.

   _____ ÷ _____ = _____

3. The radius of the circle is _____.

Example 2

What is the circumference of the circle?

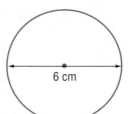
6 cm

1. The diameter is 6 centimeters; $d = 6$.

2. Substitute 6 for d and 3.14 for π in the circumference formula.

   $C = \pi d$
   $C \approx 3.14 \times 6$
   $C \approx 18.84$

The circumference of the circle is about 18.84 centimeters.

YOUR TURN!

What is the circumference of the circle?

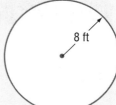
8 ft

1. The radius is _____ feet, so $r =$ _____.

2. Substitute _____ for r and 3.14 for π in the circumference formula.

   $C = 2\pi r$
   $C \approx 2 \times 3.14 \times$ _____
   $C \approx$ _____

> When you know the radius, use $C = 2\pi r$.

The circumference of the circle is about _____ feet.

Example 3

What is the area of the circle?

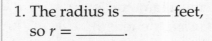
10 yd

1. The radius is 10 yards; $r = 10$.

2. Substitute 10 for r and 3.14 for π in the area formula.

   $A = \pi r^2$

   $A \approx 3.14 \times 10^2$

   $A \approx 3.14 \times 100$

   $A \approx 314$

The area of the circle is about 314 square yards.

YOUR TURN!

What is the area of the circle?

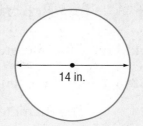

14 in.

1. Find the length of the radius.

   The diameter is _____ inches. $r =$ _____ $\div\, 2 =$ _____.

2. Substitute _____ for r and 3.14 for π in the area formula.

   $A = \pi r^2$

   $A \approx 3.14 \times$ _____^{2}

   $A \approx 3.14 \times$ _____

   $A \approx$ _____

The area of the circle is about _____ square inches.

Who is Correct?

What is the circumference of a circle with a radius of 5 yards?

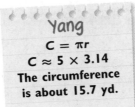

Tavio

$C = 2\pi r$

$C \approx 2 \times 5 \times 3.14$

The circumference
is about 31.4 yd.

Yang

$C = \pi r$

$C \approx 5 \times 3.14$

The circumference
is about 15.7 yd.

McKenzie

$C = \pi d$

$C \approx 10 \times 3.14$

The circumference
is about 31.4 yd.

Circle correct answer(s). Cross out incorrect answer(s).

 Guided Practice

Identify the length of the radius and the diameter of each circle.

I
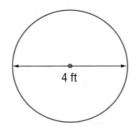

4 ft

radius: _____ ft
diameter: _____ ft

2
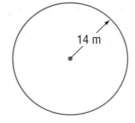

14 m

radius: _____ m
diameter: _____ m

3
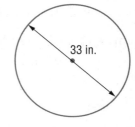

33 in.

radius: _____
diameter: _____

GO ON

4 Find the circumference and area of the circle.

> **Step 1** Find the length of the diameter and radius.
>
> The diameter is _____ centimeters.
>
> $r =$ _____ $\div 2 =$ _____.

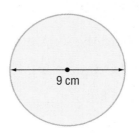

9 cm

> **Step 2** For circumference, substitute _____ for d and 3.14 for π in the circumference formula.

$C = \pi d$
$C \approx 3.14 \times$ _____
$C \approx$ _____

> **Step 3** For area, substitute _____ for r and 3.14 for π in the area formula.

$A = \pi r^2$
$A \approx 3.14 \times$ _____^{2}
$A \approx 3.14 \times$ _____
$A \approx$ _____

> The circumference of the circle is about
>
> _____ centimeters, and the area of the circle
>
> is about _____ square centimeters.

Find the circumference and area of each circle. Use 3.14 for π.

5 $d =$ _____ in.

$C = \pi d$

$C \approx 3.14 \times$ _____

$C \approx$ _____

$r =$ _____ in.

$A = \pi r^2$

$A \approx 3.14 \times$ _____^{2}

$A \approx 3.14 \times$ _____

$A \approx$ _____

13 in.

The circumference of the circle is about _____ inches, and

the area of the circle is about _____ square inches.

6 The circumference of the circle is about _____ meters, and

the area of the circle is about _____ square meters.

$d =$ _____ m

$C = \pi d$

$C \approx 3.14 \times$ _____

$C \approx$ _____

$r =$ _____ m

$A = \pi r^2$

$A \approx 3.14 \times$ _____^{2}

$A \approx 3.14 \times$ _____

$A \approx$ _____

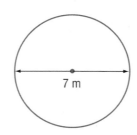

7 m

Find the circumference and area of each circle. Use 3.14 for π.

7 The circumference of the circle is about _____ yards, and the area of the circle is about _____ square yards.

6 yd

8 The circumference of the circle is about _____ feet, and the area of the circle is about _____ square feet.

4 ft

9 The circumference of the circle is about _____ inches, and the area of the circle is about _____ square inches.

11 in.

10 The circumference of the circle is about _____ centimeters, and the area of the circle is about _____ square centimeters.

17 cm

Step by Step Problem-Solving Practice

Solve.

11 **CARS** Octavia's tires each have a diameter of 35 inches. What is the circumference of each tire?

Problem-Solving Strategies
☐ Draw a diagram.
☑ Use a formula.
☐ Guess and check.
☐ Act it out.
☐ Solve a simpler problem.

Understand Read the problem. Write what you know.
Alana's tires have a diameter of _____ inches.

Plan Pick a strategy. One strategy is to use a formula. Use the formula for the circumference of a circle.

Solve Substitute _____ for d and 3.14 for $π$ in the circumference formula.

$$C = πd$$
$$C ≈ 3.14 × \_\_\_\_\_$$
$$C ≈ \_\_\_\_\_$$

The circumference of each tire is about _____ inches.

Check Estimate the circumference by substituting 3 for $π$ and 35 for d.

$$C = πd$$
$$C ≈ 3 × \_\_\_\_\_$$
$$C ≈ \_\_\_\_\_$$

The circumference of each tire is close to the estimate, so the answer is reasonable.

GO ON

12 SPORTS The skating rink has a radius of 8 yards. What is the circumference of the skating rink? Use 3.14 for π. Check off each step.

_____ Understand: I underlined key words.

_____ Plan: To solve the problem I will _____.

_____ Solve: The answer is _____.

_____ Check: I checked my answer by _____

_____.

13 HOBBIES Darin collects clocks. The face of his largest clock has a radius of 23 inches. What is the area of Darin's largest clock? Use 3.14 for π. _____

14 Reflect Using an appropriate measuring tool and the centimeter grid shown, find the circumference and area using the formulas $C = 2\pi r$ and $A = \pi r^2$. Write down both sets of numbers. How do the two sets of numbers compare?

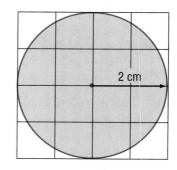

2 cm

_____

_____

_____

_____

▶ **Skills, Concepts, and Problem-Solving**

Identify the length of the radius and diameter of each circle.

15
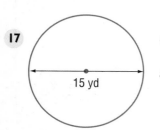
50 in.

radius = _____ in.
diameter = _____ in.

16
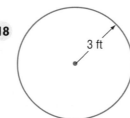
90 km

radius = _____ km
diameter = _____ km

Find the circumference and area of the circle. Use 3.14 for π.

17
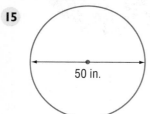
15 yd

circumference ≈ _____ yd
area ≈ _____ yd²

18
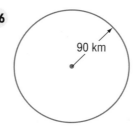
3 ft

circumference ≈ _____ ft
area ≈ _____ ft²

228 **Chapter 5 Two-Dimensional Figures**

Solve.

19 **HOBBIES**   Dane's grandmother used a circular canvas last week while painting a collage. The canvas had a diameter of 14 inches. What was the area of the canvas? Use 3.14 for π.

_____

Vocabulary Check   **Write the vocabulary word that completes each sentence.**

20 _____ is the distance around a circle.

21 _____ is the ratio of the circumference of a circle to the diameter of the same circle. Its value is approximately 3.14.

22 **Writing in Math**   Explain how to find the area of a circle.

_____

_____

 Spiral Review

Name each figure by its angles. (Lesson 5-2, p. 187)

23
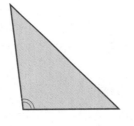

_____

24
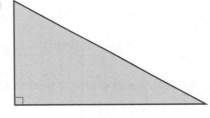

_____

Solve. (Lesson 5-5, p. 209)

25 **SAILBOATS**   The traditional boats found on Lake Tai in China have sails shaped like parallelograms. Refer to the photo caption at the right. What is the area of the sail?

_____

SAILBOATS The height of the large sail is 25 feet. The base is 16 feet.

Find the area of each parallelogram.

1
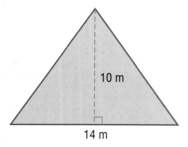
7 in.

13 in.

$A =$ _____

2
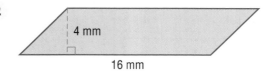
4 mm

16 mm

$A =$ _____

Find the area of each triangle.

3

10 m

14 m

$A =$ _____

4

4 ft

11 ft

$A =$ _____

Find the circumference and area of each circle. Use 3.14 for π.

5 The circumference of the circle is about _____ yards,

and the area of the circle is about _____ square yards.

8 yd

6 The circumference of the circle is about _____ feet.

The area of the circle is about _____ square feet.

12 ft

Solve.

7 **ARCHITECTURE** The owners of this house would like to paint the front and need to know how much paint to buy. What is the area if it has a height of 25 feet and a base of 40 feet?

_____

8 **BASEBALL** According to baseball regulations, the pitcher's mound, which is circular, must have a diameter of 18 feet. What is the approximate area of a pitcher's mound?

_____

Vocabulary and Concept Check

acute angle, p. 187

circumference, p. 223

obtuse angle, p. 187

pi (π), p. 223

radius, p. 223

rhombus, p. 180

square unit, p. 195

trapezoid, p. 180

Write the vocabulary word that completes each sentence.

1 _____ is the distance around a circle.

2 An angle that measures greater than 90° but less than 180° is called a(n) _____.

3 The unit used for measuring area is called a(n) _____.

4 A(n) _____ has only one pair of opposite sides parallel.

5 The distance from the center of a circle to any point on the circle is the _____.

Identify the correct figure for each formula.

6 $A = \pi r^2$

_____

7 $A = \ell \times w$

_____

8 $A = b \times h$

_____

9 $A = \frac{1}{2} \times b \times h$

_____

Lesson Review

5-1 Quadrilaterals (pp. 180–186)

Identify each figure.

10 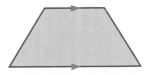   _____

11

_____

Example 1

Identify the figure.

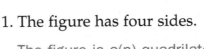

1. The figure has four sides.

   The figure is a(n) quadrilateral.

2. There are two pairs of parallel sides.

3. There are no right angles.

   The figure is a(n) rhombus.

5-2 Triangles (pp. 187–193)

Name each triangle by its sides or angles.

12

_____

13
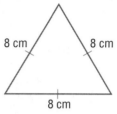

8 cm   8 cm

8 cm

_____

Example 2

Name the triangle by its sides.

6 in.   6 in.

5 in.

1. The triangle has 2 congruent sides.

2. The triangle is an isosceles triangle.

5-3 Introduction to Area (pp. 195–200)

Find the area of each figure.

14
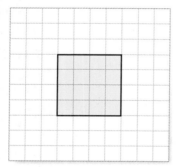

The area of the square is _____ square units.

15
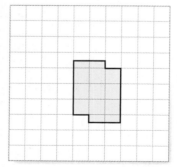

The area of the figure is about _____ square units.

Example 3

Find the area of the rectangle.

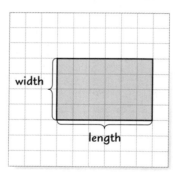

width

length

1. Count the squares the rectangle covers.

2. The area of the rectangle is **24** square units.

3. Check your answer.
   Count the squares in the top row. **6**
   Count the number of rows. **5**

4. Add the rows to find the area.

   6 + 6 + 6 + 6 = 24

The area of the rectangle is about 24 square units.

5-4 Area of a Rectangle (pp. 201–207)

Find the area of each rectangle.

16

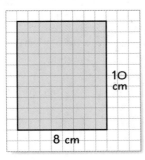

A = _____

17

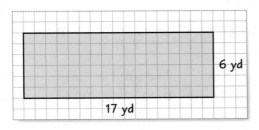

A = _____

Find the area of each square.

18

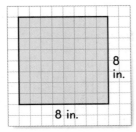

A = _____

19

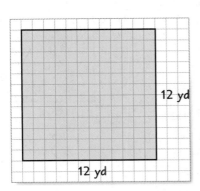

A = _____

Example 4

What is the area of the rectangle?

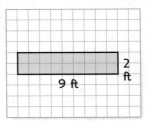

1. The length of the rectangle is 9 feet, and the width is 2 feet.

2. Substitute these values into the formula. Multiply.

   $A = \ell \times w$
   $A = 9 \text{ ft} \times 2 \text{ ft}$
   $A = 18 \text{ ft}^2$

The area of the rectangle is 18 square feet.

Example 5

What is the area of the square?

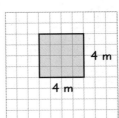

1. The length of the square is 4 meters, and the width is 4 meters.

2. Substitute these values into the formula. Multiply.

   $A = \ell \times w$
   $A = 4 \text{ m} \times 4 \text{ m}$
   $A = 16 \text{ m}^2$

The area of the square is 16 square meters.

5-5 Area of a Parallelogram

(pp. 209–214)

Find the area of each parallelogram.

20

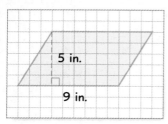

$A =$ _____

21

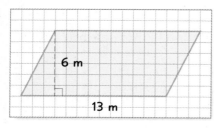

$A =$ _____

Example 6

Find the area of the parallelogram.

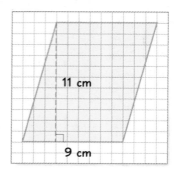

1. The base is 9 centimeters. The height is 11 centimeters.

2. Substitute these values into the formula. Multiply.

   $A = b \times h$ The area of the
   $A = 9 \text{ cm} \times 11 \text{ cm}$ parallelogram is
   $A = 99 \text{ cm}^2$ 99 square centimeters.

5-6 Area of a Triangle (pp. 215–222)

Find the area of each triangle.

22

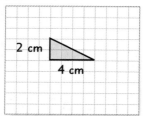

$A =$ _____

23

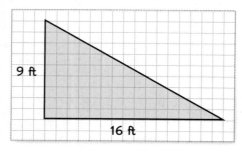

$A =$ _____

Example 7

Find the area of the triangle.

1. The base is 8 yards long.

2. The height is 17 yards long.

3. Substitute these values into the formula.

   $A = \frac{1}{2} \times b \times h$

   $A = \frac{1}{2} \times 8 \text{ yd} \times 17 \text{ yd}$

4. Multiply to find the area of the triangle.

   $A = 68 \text{ yd}^2$

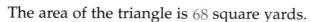

The area of the triangle is 68 square yards.

5-7 Circles (pp. 223-229)

24 Find the circumference of the circle. Use 3.14 for π.

$C \approx \pi d$

$C \approx 3.14 \times 7$

$C \approx$ _____

7 ft

25 Find the circumference of the circle. Use 3.14 for π.

$C \approx 2\pi r$

$C \approx 2 \times 3.14 \times 5$

$C \approx$ _____

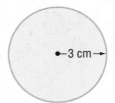

5 yd

26 What is the area of the circle? Use 3.14 for π.

$A \approx 3.14 \times 3^2$

$A \approx 3.14 \times 9$

$A \approx$ _____

3 cm

27 What is the area of the circle? Use 3.14 for π.

$A \approx 3.14 \times 5^2$

$A \approx 3.14 \times 25$

$A \approx$ _____

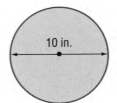

10 in.

Example 8

What is the circumference of the circle?

The diameter is 12 meters; $d = 12$.

Substitute 12 for d and 3.14 for π in the circumference formula.

$C \approx \pi d$

$C \approx 3.14 \times 12$

$C \approx 37.68$

12 m

The circumference of the circle is about 37.68 meters.

Example 9

What is the area of the circle?

1. The radius is 9 ft; $r = 9$.

2. Substitute 9 for r and 3.14 for π in the area formula.

$A = \pi r^2$

$A \approx 3.14 \times 9^2$

$A \approx 3.14 \times 81$

$A \approx 254.34$

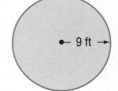

9 ft

The area of the circle is about 254.34 ft².

Identify each figure.

1

_____

2

_____

Name each triangle by its sides or angles.

3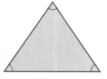

_____

4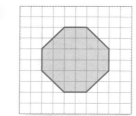

_____

Find the area of each figure.

5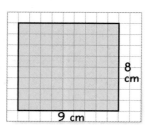

_____

6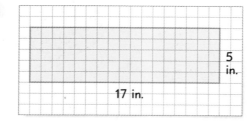

_____

Find the area of each rectangle.

7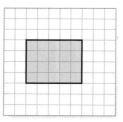

$A = $ _____

8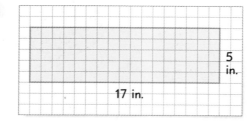

$A = $ _____

Find the area of each parallelogram.

9 $A =$ _____

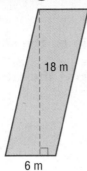

18 m

6 m

10 $A =$ _____

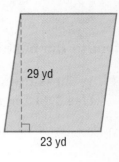

29 yd

23 yd

Find the area of each triangle.

11 $A =$ _____

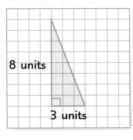

8 units

3 units

12 $A =$ _____

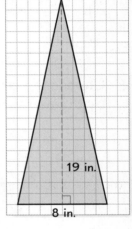

19 in.

8 in.

Find the circumference and area of the circle. Use 3.14 for π.

13 The circumference of the circle is about _____,

and the area of the circle is about _____.

16 cm

Solve.

14 **FLAGS** Lakita hung a flag shaped like a parallelogram in her bedroom. The flag is 60 inches wide at the base and 36 inches tall. What is the area of her flag? _____

15 **MEASUREMENT** Steve uses a measuring tool to measure angles of wood while building a deck. The measuring tool is shaped like a triangle. The base is 20 millimeters and it is 26 millimeters tall. What is the area of the measuring tool? _____

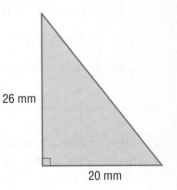

26 mm

20 mm

Correct the mistake.

16 Mia measured the area of a base of a cylinder in math class. The base of the cylinder had a radius of 6 centimeters. Mia said the area of the base of the cylinder was 18.84 square centimeters. Was she correct? Why or why not?

_____

_____

STOP

Choose the best answer and fill in the corresponding circle on the sheet at right.

1 What is the circumference of the circle? Use 3.14 for π.

12 yd

A 18.84 yd   **C** 75.36 yd

B 37.68 yd   **D** 452.16 yd

2 Choose the correct name of the triangle by the measure of its angles.

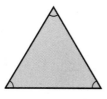

A acute triangle

B obtuse triangle

C right triangle

D none of the above

3 Devon has a parallelogram-shaped wallet. It has a base length of 9 cm and a height of 8 cm. What is the area of the wallet?

A 17 cm²   **C** 90 cm²

B 72 cm²   **D** 144 cm²

4 What is the area of the triangle?

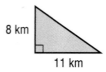

8 km

11 km

A 19 km   **C** 44 km²

B 44 km   **D** 88 km²

5 What is the area of the figure?

A 11 square units   **C** 28 square units

B 21 square units   **D** 35 square units

6 What is the area of the parallelogram?

7 km

6 km

A 13 km²   **C** 36 km²

B 30 km²   **D** 42 km²

7 Elliott's 9th grade class painted a wall of the gymnasium. The wall measures 10 feet by 36 feet. What is the area of this wall?

A 92 ft²   **C** 360 ft²

B 46 ft²   **D** 3,600 ft²

8 What is the area of the triangle?

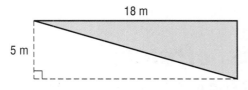

A 23 m²   **C** 90 m²

B 45 m²   **D** 180 m²

---

9 Which of the following shows a right scalene triangle?

A
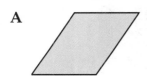

C
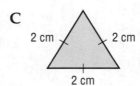

B
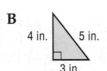

D
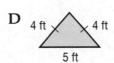

---

10 What is the area of the circle? Use 3.14 for π.

A 25.12 cm²   **C** 67.14 cm²

B 50.24 cm²   **D** 200.96 cm²

11 Choose the correct name of the figure.

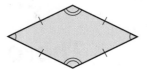

A rectangle   **C** rhombus

B triangle   **D** trapezoid

ANSWER SHEET

Directions: Fill in the circle of each correct answer.

1	Ⓐ	Ⓑ	Ⓒ	Ⓓ
2	Ⓐ	Ⓑ	Ⓒ	Ⓓ
3	Ⓐ	Ⓑ	Ⓒ	Ⓓ
4	Ⓐ	Ⓑ	Ⓒ	Ⓓ
5	Ⓐ	Ⓑ	Ⓒ	Ⓓ
6	Ⓐ	Ⓑ	Ⓒ	Ⓓ
7	Ⓐ	Ⓑ	Ⓒ	Ⓓ
8	Ⓐ	Ⓑ	Ⓒ	Ⓓ
9	Ⓐ	Ⓑ	Ⓒ	Ⓓ
10	Ⓐ	Ⓑ	Ⓒ	Ⓓ
11	Ⓐ	Ⓑ	Ⓒ	Ⓓ

Success Strategy

Read the entire question before looking at the answer choices. Make sure you know what the question is asking.

STOP

Three-Dimensional Figures

There are triangles, circles, rectangles, and other figures all around us.

You see figures every day. Everywhere you go there are different figures. Some may be flat, or solid. Shrubs can be formed into different three-dimensional figures.

Math Online Are you ready for Chapter 6? Take the Online Readiness Quiz at *glencoe.com* to find out.

STEP **2** Preview

Get ready for Chapter 6. Review these skills and compare them with what you will learn in this chapter.

What You Know	What You Will Learn
You know that two-dimensional figures are flat. 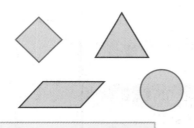	*Lesson 6-1* Three-dimensional figures are not flat. 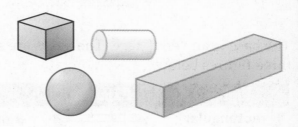
You know that if you filled a shoe box with cubes, you could count how many cubes fit into the shoe box. You also know that if you covered all of the faces of the shoe box with cubes, you could count how many cubes cover the shoe box.	*Lessons 6-2 and 6-3* Volume: number of cubes inside $V = \ell \times w \times h$ 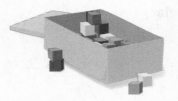 Surface Area: the sum of the number of cubes that cover all of the faces of the figure 

GO ON

Introduction to Three-Dimensional Figures

KEY Concept

Three-dimensional figures are named by the types of surfaces they have. Their surfaces can be curved, flat, or both.

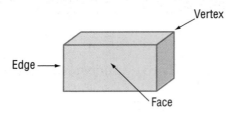

The base is an "end" face. The bases are shaded in the first three figures below.

Figure	Example	Description
rectangular prism		a prism with six rectangular faces
cube		a prism with six faces that are congruent squares
triangular prism		a prism that has triangular bases
cone		a solid that has a circular base and one curved surface from the base to a vertex
cylinder		a solid with two parallel, congruent, circular bases; a curved surface connects the bases
sphere		a solid figure that is a set of all points that are the same distance from the center

VOCABULARY

congruent figures
figures having the same size and the same shape

edge
the line segment where two faces of a three-dimensional figure meet

face
the flat side of a three-dimensional figure

three-dimensional figure
a solid figure that has length, width, and height

vertex
the point on a three-dimensional figure where three or more edges meet

Example 1

Find the number of faces, vertices, and edges. Identify the figure.

1. There are 5 faces.

2. There are 6 vertices.

3. There are 9 edges.

4. This figure is a triangular prism.

YOUR TURN!

Find the number of faces, vertices, and edges. Identify the figure.

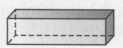

1. There are _____ faces.

2. There are _____ vertices.

3. There are _____ edges.

4. This figure is a _____.

Example 2

Identify the three-dimensional figure.

1. Is the figure flat, curved, or both? both

2. Describe the base(s).
   two congruent circular bases

3. The figure is a cylinder.

YOUR TURN!

Identify the three-dimensional figure.

1. Is the figure flat, curved, or both? _____

2. Describe the base(s).
   _____

3. The figure is a(n) _____.

Who is Correct?

Identify the three-dimensional figure.

Carissa
The figure is flat and curved. There is one circular base. The figure is a cone.

Collin
The figure is flat and curved. There are two circular bases. The figure is a cylinder.

Jade
The figure is flat. There is one circular base. The figure is a cylinder.

Circle correct answer(s). Cross out incorrect answer(s).

▶ Guided Practice

Find the number of faces, vertices, and edges.

1. Count the faces. There are _____ faces.
   Count the vertices. There are _____ vertices.
   Count the edges. There are _____ edges.

Step by Step Practice

2. Identify the three-dimensional figure.

   Step 1 Is the figure flat, curved, or both? _____

   Step 2 Describe the base(s). _____

   Step 3 The figure is a(n) _____.

Identify each three-dimensional figure.

3.

   The figure is _____.

   Describe the base(s). _____
   _____

   The figure is a(n) _____.

4.

   The figure is _____.

   Describe the base(s). _____
   _____

   The figure is a(n) _____.

5.

   The figure is _____.

   Describe the base(s). _____
   _____

   The figure is a(n) _____.

6.

   The figure is _____.

   Describe the base(s). _____
   _____

   The figure is a(n) _____.

Step by Step **Problem-Solving Practice**

Solve.

7 **PARTY**   Dante served a wedge of cheese at his party. The block of cheese is a three-dimensional figure. What is the name of the figure?

Understand   Read the problem. Write what you know.
Dante served a wedge of cheese in the shape of a

_____.

Plan   Pick a strategy. One strategy is to use a diagram.

Solve   Look at the photo. Describe the shape.

_____

_____

_____

The three-dimensional figure is a _____.

Check   Compare the figure to other figures in the lesson.

8 **HISTORY**   Samantha's 8th-grade class visited the Lincoln Memorial in Washington D.C. The columns of the structure are three-dimensional figures. What is the name of the three-dimensional figures? Check off each step.

_____ Understand: I underlined key words.

_____ Plan: To solve the problem, I will _____.

_____ Solve: The answer is _____.

_____ Check: I will check my answer by _____.

9 **DINNER**   Benito's family had spaghetti and meatballs for dinner. He wanted to identify the shape of the meatballs. What figure are the meatballs? _____

10 **Reflect**   What is the difference between three-dimensional figures and two-dimensional figures? Explain. Give an example of each.

_____

_____

_____

GO ON

Skills, Concepts, and Problem Solving

Identify each three-dimensional figure.

11

_____

12

_____

13

_____

14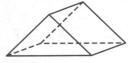

_____

15

_____

16

_____

Read each description. Identify the three-dimensional figure.

17 This figure has six rectangular faces. _____

18 This figure has a curved surface, a circled-shaped top, and
a circular-shaped base.

_____

Vocabulary Check   **Write the vocabulary word that completes each sentence.**

19 A(n) _____ is the line segment where
two faces of a three-dimensional figure meet.

20 A figure that has length, width, and height is a(n)

_____.

21 **Writing in Math**   Explain why figures with curved surfaces cannot
be prisms.

_____

_____

_____

Lesson 6-2

Surface Area of Rectangular Solids

KEY Concept

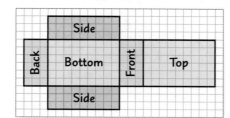

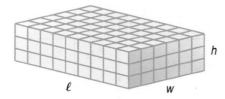

The **net** can be folded to make a rectangular prism.

The **surface area** of a rectangular prism is the sum of the areas of all the **faces** of the figure. A rectangular prism has six faces.

The following formula can be used to find surface area:

$$S = (\ell \times w) + (\ell \times h) + (w \times h) + (\ell \times w) + (\ell \times h) + (w \times h)$$

VOCABULARY

face
the flat part of a three-dimensional figure

net
a two-dimensional figure that can be used to build a three-dimensional figure

surface area
the sum of the areas of all the surfaces (faces) of a three-dimensional figure

square unit
a unit for measuring area

Example 1

Find the surface area of the rectangular prism.

1. Use a net of the rectangular prism.

2. Find the area of faces A and F.

   $A = \ell \times w$
   $A = 6 \times 5 = 30$

3. Find the area of faces B and D.

   $A = \ell \times w$
   $A = 5 \times 2 = 10$

4. Find the area of faces C and E.

   $A = \ell \times w \qquad A = 6 \times 2 = 12$

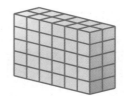

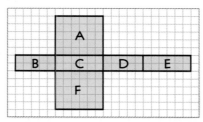

5. Find the sum of all the areas of all the faces. $30 + 10 + 12 + 30 + 10 + 12 = 104$

The surface area of the rectangular prism is 104 square units.

GO ON

YOUR TURN!

Find the surface area of the rectangular prism.

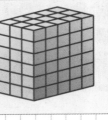

1. Use a net of the rectangular prism.

2. Find the area of faces A and F.

   $A = \ell \times w$

   $A = \underline{\hspace{1cm}} \times \underline{\hspace{1cm}} = \underline{\hspace{1cm}}$

3. Find the area of faces B and D.

   $A = \ell \times w$

   $A = \underline{\hspace{1cm}} \times \underline{\hspace{1cm}} = \underline{\hspace{1cm}}$

4. Find the area of faces C and E.

   $A = \ell \times w$

   $A = \underline{\hspace{1cm}} \times \underline{\hspace{1cm}} = \underline{\hspace{1cm}}$

5. Find the sum of the areas of all the faces.

   $\underline{\hspace{1cm}} + \underline{\hspace{1cm}} + \underline{\hspace{1cm}} + \underline{\hspace{1cm}} + \underline{\hspace{1cm}} + \underline{\hspace{1cm}} = \underline{\hspace{1cm}}$

The surface area of the rectangular prism is _____ square units.

Example 2

Find the surface area of the cube.

1. Find the area of each face.

   $A = \ell \times w$

   $A = 3 \times 3 = 9$

2. There are six faces on the cube. Find the sum of the areas of all six faces.

   $9 + 9 + 9 + 9 + 9 + 9 = 54$

The surface area of the cube is 54 square units.

YOUR TURN!

Find the surface area of the cube.

1. Find the area of each face.

   $A = \ell \times w$

   $A = \underline{\hspace{1cm}} \times \underline{\hspace{1cm}} = \underline{\hspace{1cm}}$

2. Find the sum of the areas of all six faces.

   $\underline{\hspace{1cm}} + \underline{\hspace{1cm}} + \underline{\hspace{1cm}} + \underline{\hspace{1cm}} + $
   $\underline{\hspace{1cm}} + \underline{\hspace{1cm}} = \underline{\hspace{1cm}}$

The surface area of the cube is _____ square units.

Who is Correct?

Find the surface area of the rectangular prism.

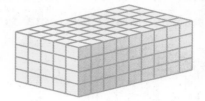

Mark

$A = 5 \times 5 = 25$ units^2

$A = 4 \times 4 = 25$ units^2

$A = 9 \times 9 = 81$ units^2

$A = 25 + 16 + 81$

$= 122$ units^2

Arturo

$A = 5 \times 9 = 35$ units^2

$A = 5 \times 4 = 20$ units^2

$A = 4 \times 9 = 36$ units^2

$A = 45 + 20 + 36 +$
$45 + 20 + 36$
$= 182$ units^2

Jarvis

$A = 5 \times 9 = 45$ units^2

$A = 9 \times 4 = 36$ units^2

$A = 5 \times 4 = 20$ units^2

$A = 45 + 36 + 20$
$+ 45 + 36 + 20$
$= 202$ units^2

Circle correct answer(s). Cross out incorrect answer(s).

▶ Guided Practice

Draw a net for a rectangular prism with the given length, width, and height.

1 $7 \times 8 \times 2$

2 $3 \times 4 \times 2$

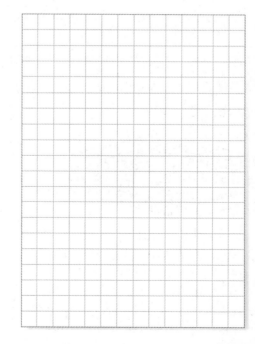

GO ON

Step by Step Practice

Find the surface area of the rectangular prism.

3

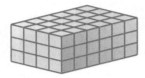

Step 1 Use a net of the rectangular prism.

Step 2 Find the area of faces A and F.

$A = \ell \times w$

$A = \underline{\hspace{1cm}} \times \underline{\hspace{1cm}} = \underline{\hspace{1cm}}$

Step 3 Find the area of faces B and D.

$A = \ell \times w$

$A = \underline{\hspace{1cm}} \times \underline{\hspace{1cm}} = \underline{\hspace{1cm}}$

Step 4 Find the area of faces C and E.

$A = \ell \times w$

$A = \underline{\hspace{1cm}} \times \underline{\hspace{1cm}} = \underline{\hspace{1cm}}$

Step 5 Find the sum of the areas of all the faces.

$\underline{\hspace{1cm}} + \underline{\hspace{1cm}} + \underline{\hspace{1cm}} + \underline{\hspace{1cm}} + \underline{\hspace{1cm}} + \underline{\hspace{1cm}} = \underline{\hspace{1cm}}$

The surface area of the rectangular prism is \underline{\hspace{1cm}} square units.

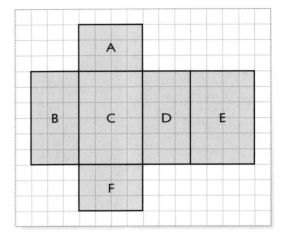

4 Find the surface area of the rectangular prism.

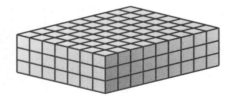

Use a net of the rectangular prism. Follow the steps at the top of page 251 to find the surface area. The surface area of the rectangular prism is \underline{\hspace{2cm}} square units.

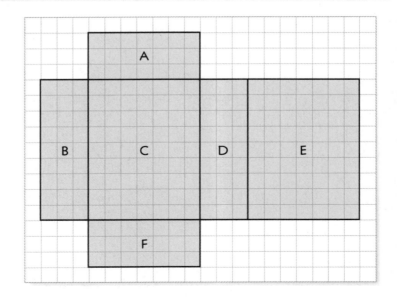

Find the area of faces A and F.

$A = \ell \times w$

$A =$ _____ $\times$ _____ $=$ _____

Find the area of faces B and D.

$A = \ell \times w$

$A =$ _____ $\times$ _____ $=$ _____

Find the area of faces C and E.

$A = \ell \times w$

$A =$ _____ $\times$ _____ $=$ _____

Find the sum of the area of all the faces.

_____ $+$ _____ $+$ _____ $+$ _____ $+$ _____ $+$ _____ $=$ _____
   A   B   C   D   E   F

The surface area of the rectangular prism is _____ square units.

Find the surface area of the rectangular prism.

5 Find the area of faces A and F.

$A = \ell \times w$

$A =$ _____ $\times$ _____ $=$ _____

Find the area of faces B and D.

$A = \ell \times w$

$A =$ _____ $\times$ _____ $=$ _____

Find the area of faces C and E.

$A = \ell \times w$

$A =$ _____ $\times$ _____ $=$ _____

Find the sum of the area of all the faces.

_____ $+$ _____ $+$ _____ $+$ _____ $+$ _____ $+$ _____ $=$ _____

The surface area of the rectangular prism is _____ square units.

6 The surface area of the rectangular prism is _____ square units.

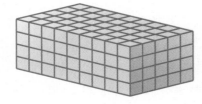

7 The surface area of the rectangular prism is _____ square units.

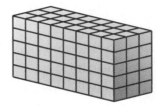

GO ON

Solve.

8 TISSUES Megan bought a box of tissues shaped like a cube. Each side measures 6 inches. What is the surface area of the box of tissues?

Understand	Read the problem. Write what you know. Each side of the box of tissues is _____ inches.
Plan	Pick a strategy. One strategy is to draw a diagram. Use a net of the cube.

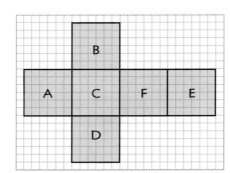

Solve   Find the area of each face.

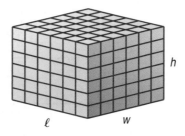

$A = \ell \times w$

$A = $ _____ $\times$ _____ $=$ _____

Find the sum of the areas of all the faces.

_____ + _____ + _____ + _____ +

_____ + _____ = _____

The surface area of Megan's box of tissues is

_____ square inches.

Check   Use a calculator to check your multiplication and addition.

9 PETS Julieta put her pet hamster in the aquarium shown. What is the surface area of the aquarium? Check off each step.

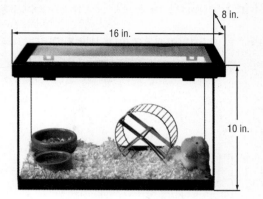

_____ Understand: I underlined the key words.

_____ Plan: To solve this problem, I will

_____ .

_____ Solve: The answer is _____ .

_____ Check: I checked my answer by _____

_____

10 GEOMETRY What is the surface area of a number cube that has 13 millimeter edges? _____

11 Reflect Use what you know about finding the area of triangles and rectangles to find the surface area of this triangular prism. (Hint: This triangular prism has 2 sides that are triangles, and 3 sides that are rectangles.)

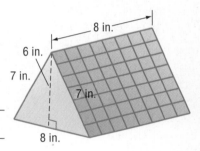

_____

_____

Skills, Concepts, and Problem Solving

Draw a net for a rectangular prism with the given length, width, and height.

12 $4 \times 4 \times 8$

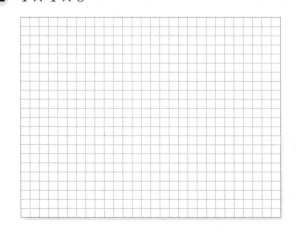

13 $3 \times 5 \times 7$

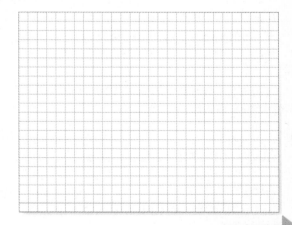

GO ON

Find the surface area of each rectangular prism.

14 The surface area of the rectangular prism is _____ square units.

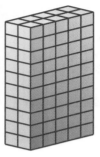

15 The surface area of the rectangular prism is _____ square units.

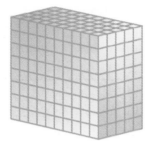

16 The surface area of the rectangular prism is _____ square units.

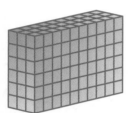

17 The surface area of the rectangular prism is _____ square units.

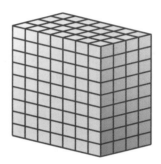

Solve.

18 **ALARM CLOCK** Juanita bought the alarm clock shown. What is the surface area of the alarm clock?

_____

ALARM CLOCK Juanita's alarm clock is shaped like a cube. The sides measure 14 centimeters each.

19 **ART** Quinn decorated a rectangular-shaped chest with wallpaper. The length of the chest is 4 feet, the width is 6 feet, and the height is 3 feet. What is the least amount of wallpaper Quinn used? _____

Vocabulary Check **Write the vocabulary word that completes each sentence.**

20 _____ is the sum of the areas of all the surfaces (faces) of a three-dimensional figure.

21 A(n) _____ is a two-dimensional figure that can be used to build a three-dimensional figure.

22 A(n) _____ is the flat side of a three-dimensional figure.

23 **Writing in Math**   Explain how to find the surface area of a rectangular prism.

_____

_____

 Spiral Review

Identify each three-dimensional figure. (Lesson 6-1, p. 242)

24

Count the faces. There are _____ faces.

Count the vertices. There are _____ vertices.

Count the edges. There are _____ edges.

25

The figure is _____.

Describe the base(s). _____

The figure is a(n) _____.

26

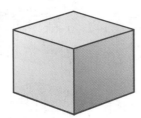

The figure is _____.

Describe the bases. _____

The figure is a(n) _____.

Solve.

27 **GEOGRAPHY**   Elena's geography class is learning how to find countries on the globe. Her teacher asks her to find Algeria. What figure is the globe? _____

28 **DONATIONS**   Rose's school is donating food to the local shelter. She brought in soup cans to donate. What figure are the soup cans she donated? _____

STOP

Identify each three-dimensional figure.

1

_____

2

_____

Draw a net for a rectangular prism with the given length, width, and height.

3 $3 \times 7 \times 6$

4 $2 \times 8 \times 5$

Find the surface area of each rectangular prism.

5 The surface area of the rectangular prism is _____ square units.

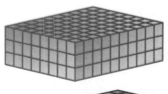

6 The surface area of the rectangular prism is _____ square units.

Solve.

7 ART Ms. Jackson asked her students to draw three-dimensional figures. She told students to use objects as models for their figures. Maurice decided to draw a cereal box. What three-dimensional figure did Maurice draw?

_____

8 STORAGE A storage cabinet is 24 inches wide, 26 inches long, and 40 inches high. What is the surface area of the storage cabinet?

_____

Introduction to Volume

KEY Concept

The amount of space inside a three-dimensional figure is the **volume** of the figure. Volume is measured in **cubic units**. To find the volume of a solid figure, determine the number of cubic units the solid figure contains.

One way to determine the volume of a **rectangular prism** is to think about the number of **cubes** in each layer.

This figure has 2 layers. Each layer has 9 cubes.

2 layers of 9 cubes = 9 + 9 = 18

This rectangular prism has a volume of 18 cubic units.

Layer 1 — 9 →
Layer 2 — 9 →

VOCABULARY

cube
a rectangular prism with six faces that are congruent squares

cubic unit
used to measure volume; tells the number of cubes of a given size it will take to fill a three-dimensional figure

rectangular prism
a three-dimensional figure that has two parallel and congruent bases in the shape of polygons; the shape of the bases tells the name of the prism

volume
the amount of space that a three-dimensional figure contains; volume is expressed in cubic units

The volume of a figure is related to its dimensions, or length, width, and height.

Example 1

Find the volume of the rectangular prism.

1. Count the number of cube layers in the prism.
   There are 2 layers of cubes.

   Layer 1 — 8 →
   Layer 2 — 8 →

2. Count the number of cubes in the top layer. There are 8 cubes in the top layer.

3. Each layer has the same number of cubes. There are 8 + 8 = 16 cubes.

The volume of the rectangular prism is 16 cubic units.

YOUR TURN!

Find the volume of the rectangular prism.

1. Count the number of cube layers in the prism.
   There are _____ layers of cubes.

   Layer 1 — 4 →
   Layer 2 — 4 →
   Layer 3 — 4 →

2. Count the number of cubes in the top layer. There are _____ cubes in the top layer.

3. Each layer has the same number of cubes. There are _____ + _____ + _____ = _____ cubes.

The volume of the rectangular prism is _____ cubic units.

GO ON

Example 2

Find the volume of the rectangular prism.

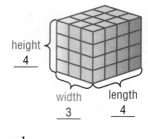

height 4

width 3

length 4

1. Look at the top layer of cubes on the prism. Finding the area of the rectangle tells you how many cubes are on that layer.

2. Area = ℓ × w, so the area of the top layer is 4 × 3 = 12. There are 12 cubes on the top layer.

3. Each layer has the same number of cubes. There are 4 layers, so there are 12 + 12 + 12 + 12 = 48 cubes in the prism.

4. The volume of the rectangular prism is 48 cubic units.

YOUR TURN!

Find the volume of the rectangular prism.

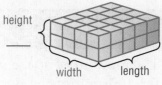

height _____

width _____ length _____

1. Look at the top layer of cubes on the prism. The length of the prism has _____ cubes. The width of the prism has _____ cubes.

2. Area = ℓ × w, so the area of the top layer is _____ × _____ = _____. There are _____ cubes on the top layer.

3. Each layer has the same number of cubes. There are _____ layers, so there are _____ + _____ = _____ cubes in the prism.

4. The volume of the rectangular prism is _____ cubic units.

Who is Correct?

Find the volume of the rectangular prism.

Abby

Each layer has 20 cubes. There are 3 layers. The volume is 60 cubic units.

Caroline

The length has 5 cubes.
The width has 4 cubes.
The height has 3 cubes.

5 + 4 + 3 = 12. The volume is 12 cubic units.

Ivan

The length has 5 cubes.
The width has 4 cubes.
There are 3 layers.
5 × 4 = 20
20 + 20 + 20 = 60
The volume is 60 cubic units.

Circle correct answer(s). Cross out incorrect answer(s).

Guided Practice

1   How many cubes are in this rectangular prism?

    _____

2   How many cubes are in this rectangular prism?

    _____

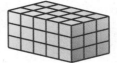

> Remember, you can find the volume of a solid figure by counting the number of cubic units it contains.

Step (by) **Step** *Practice*

3   Find the volume of the rectangular prism.

   Step 1   Count the number of cubes along the length.

   The length of the rectangular prism has _____ cubes.

   Step 2   Count the number of cubes along the width.

   The width of the rectangular prism has _____ cubes.

   Step 3   The area of the top layer is _____ × _____ = _____.

   Step 4   There are _____ layers in the prism.

   _____ + _____ + _____ = _____ cubes in the prism

   The volume of the rectangular prism is _____ cubic units.

Find the volume of each rectangular prism.

4   Count the number of cubes along the length, width, and height of the rectangular prism.

   Find the area of the top layer. Add that number four times. The volume of the rectangular prism is _____ cubic units.

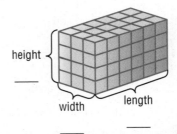

height _____

width   length

_____   _____

GO ON

5 Count the number of cubes along the length, width, and height of the rectangular prism.

Find the area of the top layer. Add that number eight times. The volume of the rectangular prism is _____ cubic units.

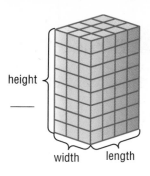

height {

____

width   length

____   ____

6    The volume of the rectangular prism is _____ cubic units.

7 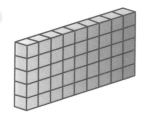   The volume of the rectangular prism is _____ cubic units.

Step by Step Problem-Solving Practice

Solve.

Problem-Solving Strategies
☐ Draw a diagram.
☐ Look for a pattern.
☑ Use a model.
☐ Solve a simpler problem.
☐ Work backward.

8 PHOTOTGRAPHY A camera that is shaped like a rectangular prism has a length of 9 centimeters, a width of 3 centimeters, and a height of 6 centimeters. What is its volume?

Understand Read the problem. Write what you know.

A rectangular prism has a length of _____ centimeters, a width of _____ centimeters, and a height of _____ centimeters.

Plan Pick a strategy. One strategy is to use a model.

Solve Use unit blocks to build the rectangular prism. Count the number of blocks used.

Check The length is _____ centimeters.

The width is _____ centimeters.

There are _____ layers.

Multiply then add.

_____ × _____ = _____

_____ + _____ + _____ + _____ + _____ + _____ = _____

The volume of the rectangular prism is _____ cubic centimeters.

9 **MODELS** Pedro's model house is 15 inches long, 20 inches wide, and 10 inches tall. What is the volume of Pedro's house?

_____

Check off each step.

_____ Understand: I underlined the key words.

_____ Plan: To solve this problem, I will _____.

_____ Solve: The answer is _____.

_____ Check: I checked my answer by _____.

10 **Reflect** Give the length, width, and height of a rectangular prism that has a volume of 36 cubic units. Explain.

_____

_____

▶ Skills, Concepts, and Problem Solving

Find the volume of each rectangular prism.

11

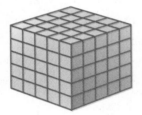

The volume of the rectangular prism is _____ cubic units.

12 The volume of the rectangular prism is

_____

cubic units.

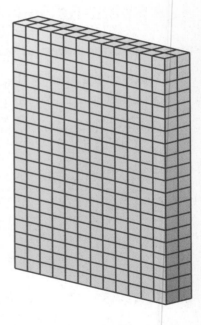

13

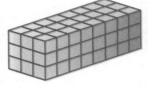

The volume of the rectangular prism is _____ cubic units.

14

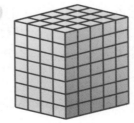

The volume of the rectangular prism is _____ cubic units.

GO ON

Solve.

15 **STORAGE** What is the volume of the storage chest shown at the right?

_____

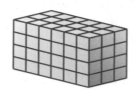

STORAGE CHEST The chest is 6 feet long, 3 feet wide, and 4 feet tall.

Vocabulary Check **Write the vocabulary word that completes each sentence.**

16 _____ is the number of cubic units needed to fill a three-dimensional figure.

17 A(n) _____ is a unit for measuring volume.

18 **Writing in Math** Explain how to find the volume of a rectangular prism.

_____

_____

▶ Spiral Review

Find the surface area of each rectangular prism. (Lesson 6-2, p. 247)

19 The surface area of the rectangular prism is _____ square units.

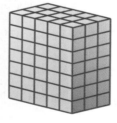

20 The surface area of the rectangular prism is _____ square units.

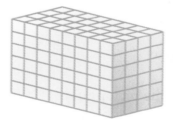

21 **SUITCASE** Tao packed his clothes in a rectangular-shaped suitcase for his trip to Florida. The length of his suitcase is 18 inches, the width is 10 inches, and the height is 12 inches. What is the surface area of his suitcase?

_____ cubic inches

12 in.

18 in.   10 in.

STOP

Volume of Rectangular Solids

KEY Concept

The amount of space inside a three-dimensional figure is the **volume** of the figure.

The volume of a rectangular solid is the product of its length, width, and height.

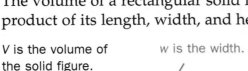

V is the volume of the solid figure.

w is the width.

$$V = \ell \times w \times h \quad \text{or} \quad V = \ell w h$$

ℓ is the length. h is the height.

VOCABULARY

cube
a rectangular prism with six faces that are congruent squares

cubic unit
used to measure volume; tells the number of cubes of a given size it will take to fill a three-dimensional figure

volume
the amount of space that a three-dimensional figure contains; volume is expressed in cubic units

Volume is measured in **cubic units**.

Example 1

Find the volume of the rectangular prism.

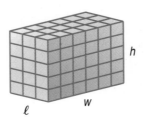

1. The length of the rectangular prism is 3 units.

   The width of the rectangular prism is 6 units.

   The height of the rectangular prism is 5 units.

2. Substitute the length, width, and height into the volume formula.

   $V = \ell \times w \times h$
   $V = 3 \times 6 \times 5$

3. Multiply.

   $V = 90$

The volume of the rectangular prism is 90 cubic units.

YOUR TURN!

Find the volume of the rectangular prism.

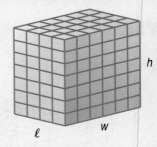

1. The length of the rectangular prism is _____ units.

   The width of the rectangular prism is _____ units.

   The height of the rectangular prism is _____ units.

2. Substitute the length, width, and height into the volume formula.

   $V = \ell \times w \times h$
   $V = \underline{\quad} \times \underline{\quad} \times \underline{\quad}$

3. Multiply.

   $V = \underline{\quad}$

The volume of the rectangular prism is _____ cubic units.

GO ON

Example 2

Find the volume of the cube.

1. The length of the cube is 6 units.

   The width of the cube is 6 units.

   The height of the cube is 6 units.

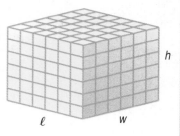

2. Substitute the length, width, and height into the volume formula.

   $V = \ell \times w \times h$
   $V = 6 \times 6 \times 6$
   $V = 6^3$

3. Multiply.

   $V = 216$

The volume of the cube is 216 cubic units.

YOUR TURN!

Find the volume of the cube.

1. The length of the cube is _____ units.

   The width of the cube is _____ units.

   The height of the cube is _____ units.

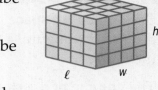

2. Substitute the length, width, and height into the volume formula.

   $V = \ell \times w \times h$
   $V = $ _____ $\times$ _____ $\times$ _____
   $V = $ _____

3. Multiply.

   $V = $ _____

The volume of the cube is _____ cubic units.

Who is Correct?

Find the volume of the rectangular prism.

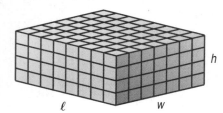

Joseph
$V = 8 \times 7 \times 4$
$= 168$ cubic units

Kerri
$V = 56 + 28 + 32$
$\quad + 56 + 28 + 32$
$= 232$ cubic units

Ramon
$V = 8 \times 7 \times 4$
$= 224$ cubic units

Circle correct answer(s). Cross out incorrect answer(s).

Guided Practice

1 How many cubes are in this rectangular prism? _____

2 How many cubes are in this rectangular prism? _____

> Check your answer. Remember, you can find the volume of a solid figure by counting the number of cubic units it contains.

Step by Step Practice

Find the volume of the rectangular prism.

3

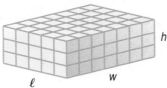

h

ℓ w

Step 1 The length of the rectangular prism is _____ units.

The width of the rectangular prism is _____ units.

The height of the rectangular prism is _____ units.

Step 2 Substitute the length, width, and height into the volume formula.

$$V = \ell \times w \times h$$

$$V = \underline{\hspace{0.6cm}} \times \underline{\hspace{0.6cm}} \times \underline{\hspace{0.6cm}}$$

Step 3 Multiply.

$$V = \underline{\hspace{0.6cm}}$$

The volume of the rectangular prism is _____ cubic units.

Find the volume of each rectangular prism.

4 Substitute the length, width, and height into the volume formula. Then multiply.

$$V = \ell \times w \times h$$

$$V = \underline{\hspace{0.6cm}} \times \underline{\hspace{0.6cm}} \times \underline{\hspace{0.6cm}}$$

$$V = \underline{\hspace{0.6cm}}$$

The volume of the rectangular prism is _____ cubic units.

GO ON

5 $V = \ell \times w \times h$

$V =$ _____ $\times$ _____ $\times$ _____

$V =$ _____

The volume of the rectangular prism is _____ cubic units.

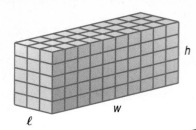

6 The volume of the rectangular prism is _____ cubic units.

7 The volume of the rectangular prism is _____ cubic units.

Step by Step Problem-Solving Practice

Solve.

8 Ruthie has a pool in her backyard that is 30 feet long, 15 feet wide, and 6 feet deep. What is the volume of Ruthie's pool?

Problem-Solving Strategies

☑ Use a model.
☐ Look for a pattern.
☐ Guess and check.
☐ Act it out.
☐ Work backward.

Understand   Read the problem. Write what you know.

The pool has a length of _____ feet, a width of _____ feet, and a height of _____ feet.

Plan   Pick a strategy. One strategy is to use a model.

Stack cubes to model the pool.

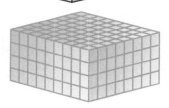

Solve   Use the formula.

$V = \ell \times w \times h$

$V =$ _____ ft $\times$ _____ ft $\times$ _____ ft

$V =$ _____ ft^3

The volume of Ruthie's pool is _____ cubic feet.

Check   Use a calculator to check your multiplication.

9 CONSTRUCTION Marion's family has a storage shed that is 5 yards wide, 8 yards long, and 5 yards high. What is the volume of the shed?
Check off each step.

_____ **Understand: I underlined the key words.**

_____ **Plan: To solve the problem, I will** _____ .

_____ **Solve: The answer is** _____ .

_____ **Check: I checked my answer by** _____ .

10 PACKAGING Mrs. Romero put together a card box for her daughter's graduation party. The card box shape is a cube. Each side measures 48 centimeters. What is the volume of the card box? _____

11 **Reflect** Compare the volume of the rectangular prism shown at the right to its surface area.

_____

_____

 # Skills, Concepts, and Problem Solving

Find the volume of each rectangular prism.

12 The volume of the rectangular prism is _____ cubic units.

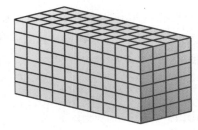

13 The volume of the rectangular prism is _____ cubic units.

14 The volume of the rectangular prism is _____ cubic units.

15 The volume of the rectangular prism is _____ cubic units.

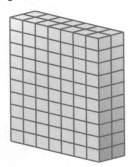

Solve.

16 FLOWER BOX   Tabitha has a flower box sitting
on her windowsill. What is the volume of the
flower box?

_____

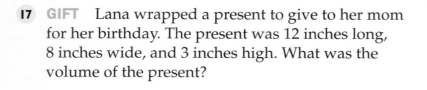

FLOWER BOX The flower box
has a length of 6 feet, width of
3 feet, and height of 3 feet.

17 GIFT   Lana wrapped a present to give to her mom
for her birthday. The present was 12 inches long,
8 inches wide, and 3 inches high. What was the
volume of the present?

_____

Vocabulary Check   **Write the vocabulary word that completes each sentence.**

18 _____ is the number of cubic units needed
to fill a three-dimensional figure.

19 _____ is a unit for measuring volume.

20 Writing in Math   Explain how to find the volume of a rectangular prism.

_____

_____

_____

▶ Spiral Review

Read each description. Identify each three-dimensional figure. (Lesson 6-1, p. 242)

21 This figure has triangular bases. _____

22 This figure has a circular base and one curved surface
from the base to a vertex. _____

Solve. (Lesson 6-2, p. 247)

23 CONSTRUCTION   Oleta's dresser is 4 feet long,
2 feet wide, and 4 feet tall. What is the surface area
of Oleta's dresser?

_____

24 MUSIC   Tobias' stereo speakers are cube-shaped. Each side
measures 20 millimeters. What is the surface area of each speaker?

_____

Progress Check 2 (Lessons 6-3 and 6-4)

Find the volume of each rectangular prism.

I 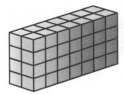 How many cubes are in this rectangular prism? _____

2 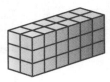 How many cubes are in this rectangular prism? _____

3 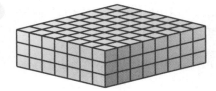 The volume of the rectangular prism is _____ cubic units.

4 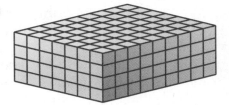 The volume of the rectangular prism is _____ cubic units.

5

The volume of the rectangular prism is _____ cubic units.

6

The volume of the rectangular prism is _____ cubic units.

Solve.

7   **PETS**   Read the caption below the photo. What is the volume of Theo's fish tank?

_____

8   **CONSTRUCTION**   Vito needed his tool box to make repairs around his house. His tool box is 36 centimeters long, 15 centimeters wide, and 8 centimeters tall. What is the volume of the tool box?

_____

9   **RECYCLING**   Gregory's recycling bin is full and ready to be picked up. The bin is 21 inches long, 14 inches wide, and 25 inches tall. What is the volume of his recycling bin?

_____

PETS Theo's fish tank is 12 inches long, 14 inches wide, and 15 inches tall.

Vocabulary and Concept Check

cube, *p. 257*

edge, *p. 242*

face, *p. 242*

net, *p. 247*

rectangular prism, *p. 257*

square unit, *p. 247*

surface area, *p. 247*

three-dimensional figure, *p. 242*

vertex, *p. 242*

volume, *p. 257*

Write the vocabulary word that completes each sentence.

1 _____ is the number of cubic units needed to fill a three-dimensional figure.

2 A(n) _____ is a two-dimensional figure that can be used to build a three-dimensional figure.

3 The sum of the areas of all the surfaces (faces) of a three-dimensional figure is the_____.

4 A(n) _____ is the line segment where two faces of a three-dimensional figure meet.

5 The point on a three-dimensional figure where three or more edges meet is the _____.

6 The _____ is the flat side of a three-dimensional figure.

7 A(n) _____ is a three-dimensional figure with six faces that are rectangles.

Label each diagram below. Write the correct vocabulary term in each blank.

8 _____

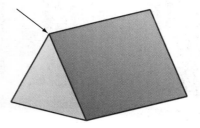

9 _____

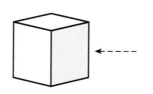

10 The net shown is of a

_____ with 6 _____.

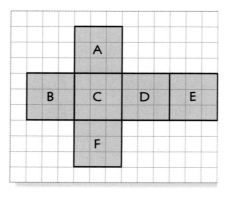

GO ON

Lesson Review

6-1 Introduction to Three-Dimensional Figures (pp. 242–246)

Identify each three-dimensional figure.

11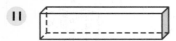

_____

12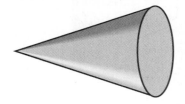

_____

Example 1

Identify the three-dimensional figure.

1. Is the shape flat, curved, or both? **both**

2. Describe the base(s). **one circular base**

3. The three-dimensional figure is a cone.

6-2 Surface Area of Rectangular Solids (pp. 247–255)

Find the surface area of each rectangular prism.

13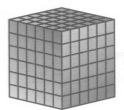

The surface area of the cube is _____ square units.

14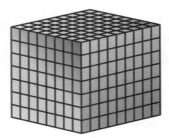

The surface area of the rectangular prism is _____ square units.

Example 2

What is the surface area of the rectangular prism?

1. Use a net of the rectangular prism.

2. Find the area of faces A and F.

   $A = \ell \times w$
   $A = 2 \times 4 = \mathbf{8}$

3. Find the area of faces B and D.

   $A = \ell \times w$
   $A = 4 \times 5 = \mathbf{20}$

4. Find the area of faces C and E.

   $A = \ell \times w$
   $A = 2 \times 5 = \mathbf{10}$

5. Find the sum of the areas of all the faces.

   $8 + 20 + 10 + 8 + 20 + 10 = \mathbf{76}$

   The surface area of the rectangular prism is 76 square units.

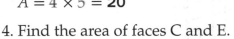

GO ON

Copyright © Glencoe/McGraw-Hill, a division of The McGraw-Hill Companies, Inc.

6-3 Introduction to Volume (pp. 257–261)

15 Find the volume of the rectangular prism.

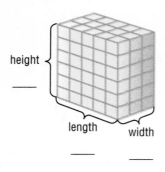

height ___

length ___   width ___

_____

_____

16 Find the volume of the rectangular prism.

_____

_____

17 Find the volume of the rectangular prism.

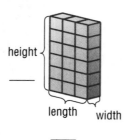

height ___

length ___   width ___

_____

_____

Example 3

Find the volume of the rectangular prism.

1. Count the number of layers of cubes in the prism. There are 5 layers of cubes in the rectangular prism.

2. Count the number of cubes in the top layer. There are 4 cubes in the top layer.

3. Each layer has the same number of cubes. There are $4 + 4 + 4 + 4 + 4$ cubes.

   The volume of the rectangular prism is 20 cubic units.

Example 4

Find the volume of the rectangular prism.

1. Look at the top layer of cubes on the prism. Finding the area of a rectangle would tell you how many cubes are on that layer.

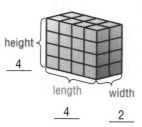

height ___ 4

length ___ 4   width ___ 2

2. Area = $\ell \times w$, so the area of the top layer is $4 \times 2 = 8$, there are 8 cubes on the top layer.

3. Each layer has the same number of cubes. There are 4 layers so there are $8 + 8 + 8 + 8 = 32$ cubes in the prism.

   The volume of the rectangular prism is 32 cubic units.

6-4 Volume of Rectangular Solids (pp. 263–268)

Find the volume of each rectangular prism.

18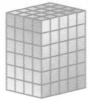

The volume of the rectangular prism is _____ cubic units.

19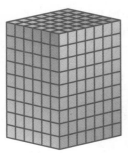

The volume of the rectangular prism is _____ cubic units.

20

The volume of the rectangular prism is _____ cubic units.

21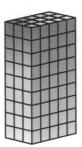

The volume of the rectangular prism is _____ cubic units.

Copyright © Glencoe/McGraw-Hill, a division of The McGraw-Hill Companies, Inc.

Example 5

Find the volume of the rectangular solid.

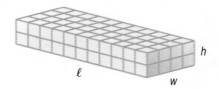

ℓ is the length.   h is the height.

$$V = \ell \times w \times h$$

V is the volume of the solid figure.   w is the width.

1. The length of the cube is 11 units.

   The width of the cube is 4 units.

   The height of the cube is 2 units.

2. Substitute the length, width, and height into the volume formula.

   $V = \ell \times w \times h$
   $V = 11 \times 4 \times 2$

3. Multiply.

   $V = 88$

   The volume of the rectangular prism is 88 cubic units.

STOP

Find the surface area of each rectangular prism.

1 The surface area of the rectangular prism is _____ square units.

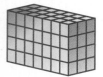

2 The surface area of the rectangular prism is _____ square units.

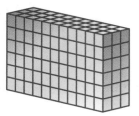

3 The surface area of the rectangular prism is _____ square units.

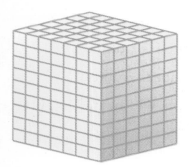

4 The surface area of the rectangular prism is _____ square units.

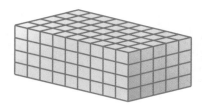

Find the volume of each rectangular prism.

5 How many cubes are in this rectangular prism? _____

6 How many cubes are in this rectangular prism? _____

7 The volume of the rectangular prism is _____ cubic units.

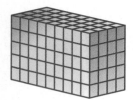

8 The volume of the rectangular prism is _____ cubic units.

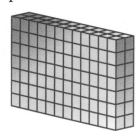

Identify each three-dimensional figure.

9

_____

10

_____

11

_____

12

_____

13

_____

14

_____

15 **MOVIES** Silvia bought a DVD player with the money she saved from baby-sitting. The length of the DVD player is 30 centimeters, the width is 25 centimeters, and the height is 4 centimeters. What is the surface area of the DVD player?

_____

16 **GARDEN** Casandra's birdhouse is 5 inches long, 4 inches wide, and 9 inches tall. What is the volume of Casandra's birdhouse?

_____

17 **MUSIC** Howard plays keyboards in the school band. His keyboard is 36 inches long, 10 inches wide, and 3 inches tall. What is the surface area of Howard's keyboard?

_____

Correct the mistakes.

18 Ms. Blackwell asked students in her math class to draw three-dimensional figures. She set some objects on a table. She told students to use the objects as models for their figures. Yolanda decided to use the roll of paper towels as a model. Look at Yolanda's drawing. What did she do wrong?

_____

_____

Choose the best answer and fill in the corresponding circle on the sheet at right.

1 What is the volume of the solid figure?

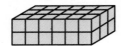

  A 12 cubic units   **C** 30 cubic units

  B 18 cubic units   **D** 36 cubic units

2 Which of the following best describes the shape of a soccer ball?

  A cone

  B rectangular prism

  C sphere

  D vertex

3 Ray has a closed shoe box that measures 10 inches by 7 inches by 5 inches. What is the volume of the shoe box?

  A 22 in³   **C** 350 in²

  B 350 in³   **D** 350 in

4 Which of the following figures shows a cone?

5 What is the surface area of the rectangular solid?

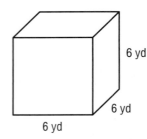

6 yd
6 yd
6 yd

  A 18 yd²

  B 186 yd²

  C 216 yd²

  D 248 yd²

6 What is the volume of the solid figure?

  A 9 cubic units   **C** 16 cubic units

  B 12 cubic units   **D** 18 cubic units

7 What is the surface are of the rectangular solid?

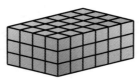

  A 108 square units

  B 72 square units

  C 54 square units

  D 13 square units

8 Patty has a jewelry box that measures 15 centimeters long by 10 centimeters wide by 8 centimeters tall. What is the volume of the jewelry box?

A 33 cm³   **C** 1,200 cm³

B 1,500 cm³   **D** 800 cm³

---

9 Choose the correct name of the three-dimensional figure.

A square   **C** rectangle

B cube   **D** box

---

10 What is the surface area of the rectangular prism?

A 47 square units   **C** 60 square units

B 12 square units   **D** 94 square units

---

11 Omar built a storage chest for his bedroom that is 5 feet long, 4 feet wide and 2 feet tall. What is the surface area of the chest?

A 47 square feet   **C** 94 square feet

B 76 square feet   **D** 141 square feet

12 Which of the following describes the object below?

A triangular prism

B sphere

C cube

D rectangular prism

ANSWER SHEET

Directions: Fill in the circle of each correct answer.

1 Ⓐ Ⓑ Ⓒ Ⓓ
2 Ⓐ Ⓑ Ⓒ Ⓓ
3 Ⓐ Ⓑ Ⓒ Ⓓ
4 Ⓐ Ⓑ Ⓒ Ⓓ
5 Ⓐ Ⓑ Ⓒ Ⓓ
6 Ⓐ Ⓑ Ⓒ Ⓓ
7 Ⓐ Ⓑ Ⓒ Ⓓ
8 Ⓐ Ⓑ Ⓒ Ⓓ
9 Ⓐ Ⓑ Ⓒ Ⓓ
10 Ⓐ Ⓑ Ⓒ Ⓓ
11 Ⓐ Ⓑ Ⓒ Ⓓ
12 Ⓐ Ⓑ Ⓒ Ⓓ

Success Strategy

After answering all the questions, go back and check your work. Make sure you circled the correct answer to each problem.

STOP